T0178320

SpringerBriefs in History of Science and Technology

More information about this series at http://www.springer.com/series/10085

Arne Schirrmacher

Establishing Quantum Physics in Göttingen

David Hilbert, Max Born, and Peter Debye in Context, 1900–1926

Springer

Arne Schirrmacher
Institut für Geschichtswissenschaften
Humboldt-Universität zu Berlin
Berlin, Germany

ISSN 2211-4564   ISSN 2211-4572   (electronic)
SpringerBriefs in History of Science and Technology
ISBN 978-3-030-22726-5   ISBN 978-3-030-22727-2   (eBook)
https://doi.org/10.1007/978-3-030-22727-2

This Springer imprint is published by the registered company Springer Nature Switzerland AG
The registered company address is: Gewerbestrasse 11, 6330 Cham, Switzerland

Preface

It might stir some irritation that this book on quantum mechanics in Göttingen does without the name of Werner Heisenberg in its title. Rather David Hilbert, the mathematician, and Peter Debye, best known for his experimental work elsewhere, are mentioned together with Max Born, the undisputed physics leader of the 1920s in the North German university town.

This contextual study of the establishing of quantum theory in Göttingen aims at providing a new perspective on a well-known and much-praised chapter of the history of science. It tries to avoid some pitfalls that often come with conceptual (re-)constructions of the emergence of quantum mechanics and widespread shortcut narratives of the main—and great—actors and their bold ideas. It needed, however, much more for a revolution of the physical description of nature to take place. From today's research conditions, it appears only too obvious that we have to talk about long-term research planning, hiring staff, securing resources and convincing both peers and the public. A hundred years ago, it was not that much different, so my basic argument, that is developed along various lines, corresponding to specific kinds of resources necessary for the creation of a new physics. Connecting the emergence of new theory (not independently from experimental work) with local conditions, management of generational change in professors and staff, ongoing discussion circles, teaching curricula as well as more general (philosophical) programmes, individual career strategies and national and international networking opens up a wider picture of a (socio-)cultural history of physics as exemplified in this case study.

From this perspective, the history of quantum theory presented here does not put the spotlight on the 'discoveries' in 1900 and 1925, when the quantum hypothesis and the first formulation of quantum mechanics were 'found', but rather attempts to shed light on the multiple, entangled and sometimes contradictory trajectories that connect the two dates and which not only cut through many layers of physical and mathematical problems, but also institutional, personal and political ones.

This work is the result of a long preoccupation with the history of quantum theory and Göttingen physics, which was pursued at a number of institutions ranging from the Research Institute of the Deutsches Museum in Munich and the

Hilbert Edition project in Göttingen to the Max Planck Institute for the History of Science and the Humboldt University both in Berlin.

John Heilbron in Berkeley and elsewhere was pivotal for the project in many ways and at different stages, while Helmuth Trischler opened space and time to launch a thorough archival research project on the history of quantum physics at the Deutsches Museum in Munich. Jürgen Renn who, after a decisive impulse already at the early stages of the project, later invited me to integrate my work into the Quantum History Project of the Max Planck Institute brought me to Berlin where I am indebted not least to the late historian Rüdiger vom Bruch, who was an invaluable interlocutor on the history of universities. From the great number of colleagues who have been a valuable help in many ways, I would like to thank in particular Finn Aaserud, Massimiliano Badino, Arianna Borrelli, Cathryn Carson, Leo Corry, Olivier Darrigol, Michael Eckert, Moritz Epple, Paul Forman, Bretislav Friedrich, Wilhelm Füßl, Ulf Hashagen, Klaus Hentschel, Dieter Hoffmann, Jeremiah James, Christian Joas, Martin Jähnert, Horst Kant, Shaul Katzir, Andreas Kleinert, Alexei Kojevnikow, Helge Kragh, Albert Krayer, Christoph Lehner, Daniela Monaldi, Jaume Navarro, Ulrich Majer, Volker Peckhaus, David Rowe, Tilman Sauer, Urs Schöpflin, Erhard Scholz, Suman Seth, Skuli Sigurdsson, Richard Staley, Renate Tobies, Mark Walker and Stefan Wolff.

Berlin, Germany   Arne Schirrmacher
December 2018

Contents

Abbreviations of Archives and Titles

AHQP	Archive for History of Quantum Physics. American Philosophical Society, Philadelphia
Bohr CW	Niels Bohr: Collected Works. 12 Vols., Amsterdam 1972–2007
DMA	Deutsches Museum, Archives
Einstein CP	Albert Einstein: Collected Papers, Princeton 1987ff
GStA PK	Geheimes Staatsarchiv Preußischer Kulturbesitz
Hilbert Papers	Niedersächsische Staats- und Universitätsbibliothek Göttingen, Special Collections, Papers of David Hilbert (Cod. Ms. D. Hilbert)
Jb. DMV	Jahresberichte der Deutschen Mathematikervereinigung
Nachr. GWG	Nachrichten von der Gesellschaft der Wissenschaften zu Göttingen
SBB	Staatsbibliothek zu Berlin Preusischer Kulturbesitz, Handschriftenabteilung
UA-Fra	Universitätsarchiv Frankfurt
UA-Gö	Universitätsarchiv Göttingen

Chapter 1
Situating Göttingen in the History of Quantum Physics: A Contextual Approach

Göttingen is without doubt one of the central places where quantum physics developed. Here quantum mechanics was established for the first time in the collaboration of Max Born with Werner Heisenberg and Pascual Jordan, two of his former students who had grown up to become real partners in science. Collecting the specific contributions to quantum theory that came from Göttingen in the two decades before the quantum mechanical revolution, however, does not combine easily into a satisfactory narrative.

Relevant publications actually emerged as isolated and mostly unrecognized works. The first were by Walter Ritz on his combination principle and by Max Abraham on black-body radiation (Ritz 1903, 1908; Abraham 1904).[1] Paul Ehrenfest, too, was quite interested in the quantum hypothesis early on, for example, he was instrumental in recognizing the inevitable discontinuity it entailed, and he also spent much time in Göttingen. However, he was not able to gain a position or influence here (Klein 1970). More prominent work had been done by Walther Nernst and Johannes Stark. Nernst developed his heat theorem while already preparing to leave for Berlin. Moreover, Stark made his discovery only after he left Göttingen, having taken with him the general idea for the Stark effect from Woldemar Voigt (Stark 1987, 22).

Born turned to Thomson's atom for his habilitation lecture and attended Albert Einstein's Salzburg talk in 1909, thus developing an interest in atomic physics and the quantum question. He contributed to the emergent research field through work with Rudolf Ladenburg on black-body radiation, with Theodore von Kármán in their papers on specific heat (at the same time as Peter Debye), and with Richard Courant relating quantum theory to the law of Eötvös (and thus surface tension of liquids),

---

[1] As Abraham did his doctorate with Planck and was his assistant from winter term 1897/98 to winter term 1900/01, he hence witnessed Planck's work culminating in his law, while Planck in turn recognized his work, cf. Planck (1906, 68).

© The Author(s), under exclusive license to Springer Nature Switzerland AG 2019
A. Schirrmacher, *Establishing Quantum Physics in Göttingen*,
SpringerBriefs in History of Science and Technology,
https://doi.org/10.1007/978-3-030-22727-2_1

as well as in a number of single-authored papers he wrote from 1911 on Born and Courant (1913). Both Born and Debye focused during the war on the question of what results would come from Bohr's atom if taken seriously. Born, rather than fully embarking on his new position as professor in Berlin a few months before the war, together with Alfred Landé and Erwin Madelung soon formed a kind of Berlin outpost of Göttingen physicists in military service and used Bohr's quantum theory for the constitution of matter, while Debye and Paul Scherrer applied it to the diffraction of X-rays. Debye also brought in the quantum for the Zeeman effect (in parallel with Arnold Sommerfeld). At this time the quantum, as the key to the atomic structure of matter, became the Göttingen credo. After the war—when the failures of the Bohr atom became apparent—it was the quantum that opened up the fruitful road to matrix mechanics as paved by Born and James Franck.

In this way, a brief sketch of the contributions to quantum theory from Göttingen can be given; however, it hardly indicates a conclusive conceptual development, nor does it refer to the conditions and driving forces behind it. One person who will be considered central in this study, the mathematician David Hilbert might not appear at all, as his publications on quantum or atomic problems are none and citations of his influence few. And such a perspective on the Göttingen story of quantum theory throws into question what the result of a local perspective should be. It might just confirm a view that the only sensible way to approach the historical development of quantum theory is a purely conceptual one that focuses on idealized discussions in journals, instead of real-life discourses, possibly enriched by some celebrated conferences and meetings. Everything else appears to spring from the ingenuity of the individual researchers. But as the only context in this approach is purely cognitive, probably consisting of personal ensembles of knowledge and world views, such a study would loose sight of the inevitably wider contexts of the creation of knowledge relating to particular local circumstances.

1.1 Local Contexts: Resources and Research Politics

To avoid this outcome, I suggest exploiting a number of indicators for scientific change in order to reconstruct the local situation in which research fields are created and altered. Understanding scientific activity as investments of resources, while de-emphasizing "pure genius" and the inherent logical structure of scientific theory, a local story of the establishment of quantum theory in Göttingen becomes feasible. I will call this scientific entrepreneurship *research politics*. It comprises decisions to direct or redirect resources in a way that favors or encumbers research in a certain field, irrespective of whether these resources are financial, personal, directed toward public perception, or even relevant for exploitable technology. In this sense, seeking for research politics is at the same time a kind of an economic or resource-oriented history of science, for it asks what resources were invested to what anticipated end. These expectations and hopes, however, often do not correspond with the successes the investments yield, a fact hard to handle from a viewpoint focused on a conceptual development. In short, I will argue for the thesis that a sound historical account of the development of quantum physics cannot properly be given without dwelling on these kinds of research politics.[2] In consequence, this study will depart to some extent from other treatments of this period, both in perspective and sources.[3]

Why rather provincial Göttingen became a place for eager advocates of quantum physics at all, and how it happened that a specific constellation was created that enabled quantum mechanics to be established here, is far from obvious. To answer this question, I go back to the local context that is out of the center of a strictly conceptual description of the historical development of quantum physics. I will analyze the relevant resources, i.e. personnel, experimental facilities, but also the accumulated knowledge—that is the cognitive resources—which might have made Göttingen a favorable place for quantum theory. In addition, I will look for an organizer or a pressure group that was driven by more or less clear objectives. This serves to determine the factors that may have turned Göttingen into a particularly favorable place with regard to the emerging and competing centers of quantum and atomic physics. Furthermore, disciplinary and institutional constraints that may have influenced the development (or that had to be overcome) are analyzed as well as the question to what extent a particular science policy by the state encouraged the reorientation.

A closer look will also be taken at how the individual researchers reacted to the news about quanta and how they adjusted their interests, both in research and teaching, towards quantum problems, and how they accepted or rejected state support and decisions, in particular with regard to filling positions. The process between the reception of the theory and participation in work on quantum problems, on fulfilling academic duties to reorganizing the curriculum, appears most important. Three key figures of the Göttingen development will be discussed in some detail: David Hilbert, Peter Debye and Max Born. They will be contrasted with the older principal Göttingen physicists Woldemar Voigt and Eduard Riecke, whom they complemented and then replaced, as well as with other colleagues.

The foremost aim of this study is to describe the various areas of research politics that can be inferred from the Göttingen story of quantum physics. However, it will then be a second step to combine this approach with a certain conceptual point of view, which brings to the fore cognitive preconditions that underlie decisions on research politics. In our case, different ways of organizing knowledge can be seen as the root of different levels of confidence in atomic reality and reductive

---

[2]The use of the term research politics, instead of *policy*, reflects the contextual point of view taken in this study in order to elucidate a development that does not exhibit a linear story. While any policy is related to a plan, a philosophy, to principles, guidelines etc., *politics* emphasizes the probably less coherent actual doings, the concrete order of steps taken, the process of trial an error etc., hence a development that may in cases lead to rather unplanned results. — The concept of resources and research politics have been developed some years ago, cp. Schirrmacher (2000, 2002, 2003).

[3]Part two on "The Emergence of the Quantum Discontinuity, 1905–1912" by Kuhn (1978) is a particularly important starting point for this study. Although the focus is on concept development and propagation, aspects like the quantitative analysis of publications suggest developing further indicators of interest, and investing time and funds, which can be localized. Cf. for the period under investigation also (Hermann 1971; Hund 1967, 1987; Mehra and Rechenberg 1982).

thinking, and hence had implications for the emergence of quantum physics. Here, the organization of knowledge refers to the ways of arranging phenomena and their description according to categories that depend on a broader scope of scientific, personal, and cultural determinants and which sometimes amount to an elaborate vision.[4] How local conditions and potentials can be combined, or rather, how they can result in a programmatic mobilisation will be shown next for Göttingen at the beginning of the 20th century.

1.2 Setting the Stage: Voigt and Hilbert

The separation of physical and mathematical research programs is part of the evolution of the sciences, and in particular of the disciplinary differentiation often seen as an indicator of scientific maturity. For these two branches of knowledge, however, a complete separation does not appear feasible. Only the major research questions can be disentangled as, roughly speaking, mathematicians seek general sets of logical schemes, while physicists are looking for the one and only completely satisfactory theory that applies to nature. If Göttingen stands for an exceptional collaboration between mathematics and physics *after* their disciplinary emancipation, this must be regarded more as an effect of a higher order than as a retrograde step.

The disciplinary situation of physics in Göttingen, as described in the seminal study by Jungnickel and McCormmach, which is taken here as a starting point, was rather special even back in 1883, since

> [...] physics was counted as mathematics [...] and the physicists Weber, Riecke, and earlier Listing, belonged to the mathematical, not the physical, class of the Göttingen Royal Society of Sciences. The practical advantage was that at Göttingen Voigt could find over 130 well-prepared, hard-working students working in mathematics.[5]

The mathematical physics institute was intended for "advanced physics through exact measurements in the laboratory" (Jungnickel and McCormmach 1986, 113). The work was divided between the two full professors:

> Riecke would retain control of the beginners' exercises and the examinations for students of medicine, pharmacy, and agriculture, but he would divide with Voigt the advanced exercises and the examinations for secondary teachers and doctoral candidates (Jungnickel and McCormmach 1986, 115).

Voigt arranged laboratory practice courses for the more mathematically trained students and paid a mathematics student to assist him (exhibiting a kind of reversal

---

[4] A further thesis to be argued for elsewhere is hence that Hilbert, on the basis of reductive thinking (axiomatics, atomism), formed the vision of an *Einheitswissenschaft* (unified science)—a conceptual scheme valid for all science—that played a crucial role in Göttingen becoming a center for quantum physics. Cp. Corry (2004), McCormmach (1987).

[5] Jungnickel and McCormmach (1986, 115), paraphrasing a letter of H. A. Schwarz to Voigt, 7 January 1883.

of roles between mathematics and physics that we will encounter later in various ways). He used his personally owned instruments for his research, and there was no doubt that his "optical work belonged to an active research area in Germany [...]"(Jungnickel and McCormmach 1986, 116).

Voigt, educated in Königsberg, had completed his dissertation in 1874 under Franz Neumann and succeeded him in the following year as an associate professor. He was eventually granted a full professorship in Göttingen in 1883 where he remained throughout his life.[6] For Voigt precise measurements were the primary task in his mathematical physics institute and theory was a good grounding for this task. The method employed was "to draw mathematical consequences from a few general principles based on experience [...] rather than to draw them from special pictures or mechanisms." "By ignoring special mechanisms, Voigt impressed upon theoretical physics a characteristic direction." This was apparent, for example, in his textbooks of the years before 1900, but it was not in terms of research facilities and buildings. A new institute had been promised upon his appointment in 1883, yet the new buildings were not inaugurated before 1906. Voigt was fairly well equipped for his main research field of optics and crystal physics, for his interest in the Zeeman effect "he had begged and borrowed spectroscopic apparatus and a powerful electromagnet [...] These were small gifts compared with what industrialists gave Göttingen's applied physics institutes [...]" (Jungnickel and McCormmach 1986, 116, 124, 269).

It was the rapidly increasing number of students, in particular, that made Voigt worry about necessary equipment, and this situation eventually rendered him more or less unable to do further research. In a ceremonial address to the university about "physics research and teaching in Germany during the last hundred years" in 1912, Voigt did not hesitate to allude to the situation at his institute:

> Of course, the red [brick] walls of the new institute tell nothing of the long, tenacious fights that preceded their erection, nothing of the budget calamity that prevails behind them, nothing of the strange means by which the operation of the institute alone is adequately maintained, nothing of the overcrowding that, true to our predictions, occurred within a few years after the institute's inauguration (Voigt 1912, 16).

The "strange" means mentioned here—viz. the ways of operation and unconventional funding—indeed point to Voigt's personal subsidies for equipment and personnel, which, according to his assistant at the time, sometimes equaled the regular budget. Too late he realized that this way of improving working conditions was wrong (Försterling 1951, 221). The point he repeatedly made, that even a provincial institute like Göttingen should have the right to be well-equipped at least in one field (like spectroscopy in his case), never did receive a warm welcome from the state.

As for the topic of the establishment of quantum physics, it might be instructive to recall Voigt's research efforts on the Zeeman effect, an important proto-quantum problem[7] that contributed greatly at various stages in the development of atomic and

---

[6]For literature on Voigt cf. Runge (1920), Föorsterling (1951), Goldberg (1970b), Jungnickel and McCormmach (1986), Heilbron (1994), Wolff (1996).

[7]The term "proto-quantum" problem was introduced by Jürgen Renn, cp. Renn (2000).

quantum physical concepts. In an address delivered on the occasion of the inauguration of the new buildings, which took place in 1906, Voigt told his audience, which included the 1902 Nobel laureate Pieter Zeeman:

> As I have dealt with the theory of these phenomena for a long time, I have also aimed to provide a home for their experimental investigation in our institute. For it is true that this physics laboratory of a provincial university naturally cannot be adequately equipped for all research fields; but it may claim to own first-rate instruments for those problems that have been creatively worked on at this place (Voigt 1906, 41).

Having made clear his dissatisfaction with the budget of his institute that was of a "depressing austerity" [bedrückende Dürftigkeit], bankruptcy was prevented only through private sacrifices. He even addressed the representatives of the Prussian Ministry of Culture directly, stating that

> One can only hope that if only the Royal state government supports us with a belated grant of a moderate fraction of the curtailed 50 percent, we will someday get a truly productive institution for the study of the Zeeman effect and its related phenomena (Voigt 1906).

Voigt's book of 1908 on magnetic and electrical optics, though motivated by Zeeman's and Lorentz's work, was later praised for its mathematical elegance and great orderliness rather than any experimental or theoretical advances (Goldberg 1970b). And Voigt himself regarded his *opus magnum* on crystal physics of 1910 as a document of estrangement from current interest (Jungnickel and McCormmach 1986, 273).

Being a good friend of both Zeeman and Lorentz, with whom he exchanged a host of letters, was one side of Voigt's scientific life,[8] the largely unsuccessful fights for resources another. By 1911 things had not changed significantly as he writes to Lorentz:

> In the inner self I always feel oppressive about how little I can accomplish, and when I look back tiredly in the evening after the day and realize with what trifling work it was spent— discussions with students on elementary theoretical and technical questions—then my job appears subaltern to me. I recently wrote to Planck: while he was moving with others—with you in particular—in the pure ether of the most general questions, I, as a mole, was digging in the soil for minor specialties.[9]

Voigt's colleague Eduard Riecke came to Göttingen on a scholarship in 1870 and obtained his doctorate under Wilhelm Weber in the following year. He became Weber's assistant and received the venia legendi for both physics and mathematics. In 1873 he became professor and he finally succeeded his teacher as a full professor in 1881.[10] Riecke and Voigt did not separate their research fields and equipment, but

---

[8] Voigt's correspondence with Lorentz and Zeeman of many years is a source rarely consulted; cf. e.g. Wolff (1996).

[9] Voigt to Lorentz, 18 August 1911. Lorentz Correspondence IV (AHQP). The last sentence reads in the original: "Ich schrieb noch kürzlich an Planck: während er mit Anderen—mit Ihnen insbesondere—sich im reinen Aether der allgemeinsten Fragen bewegte, wuhlte ich als Maulwurf in der Erde nach kleinen Spezialitäten."

[10] For literature on Riecke cf. Voigt (1915c), Sommerfeld (1916b), Goldberg (1970a), Jungnickel and McCormmach (1986).

demonstrated great harmony in sharing the institute and its facilities. It was, however, also a marriage of purpose in part, as their interests and attitudes differed in many respects. Riecke, for example, in distinction to Voigt, made great use of hypothetical models and worked on cathode rays to show their independence from the cathode material in order to confirm the existence of the electron. As one may summarize his work with Geissler tubes, on atmospheric electricity, on the behavior of crystals and on the electrical conduction of metals from 1880 to 1915 under the main objective of proving an atomic structure of electricity, an identification of his work as dealing with proto-quantum problems still does not appear straightforward.

Riecke is hence probably less central than Voigt, at least his pronounced statements and activities are few. Concerning his views on the future of the quantum ideas in particular, there is one remark of significance in a letter to Johannes Stark. Riecke wrote in October 1911—as we will see, briefly after quantum physics entered research and teaching in Göttingen—that

> [...] relativity principle and quantum theory I do not consider to be definite forms in which we can put our physical knowledge, but physics will be certainly be a good step further when it has exhausted everything that can be learned from these principles or concepts. At the beginning of the last semester, we had big but inconclusive debates about the relativity principle in the physical and the mathematical society. The mathematicians are hypnotized by the elegance of the rules of calculation, the physicists critical.[11]

In Voigt's eyes, Riecke stood for a modest and conservative research program of precise measurements similar to his own, "[...] where he performs experiments, it is usually never a question of pioneering into undeveloped fields, but of doing measurements under the guidance of theory."[12]

Apparently Riecke made peace with quantum theory only in the last month before his death, as a brief overview of Bohr's theory from his pen shows, which appeared posthumously and was immediately used to reinterpret his attitude. It was Debye who turned Riecke's strong reservations into an ostensible motivation to deal with the quantum in the first place, when he wrote in the introduction to his late colleague's final publication that he had "followed with rising interest and youthful freshness the new and far-reaching achievements physical research had gained by the extension of the sphere of influence of Planck's quantum of action, literally until his very last days."[13] The Göttingen mathematicians enjoyed a much higher visibility and reputation than their physics colleagues Riecke and Voigt. Felix Klein and David Hilbert

---

[11]Riecke to Stark, October 13, 1911, cited in Tobies (1994, 348).

[12]"Und wo er experimentiert, handelt es sich der Regel nach nicht um das pioniermäßige Vordringen in ein noch unerschlossenes Gebiet, sondern nach Messungen nach Anleitung der Theorie" (Voigt 1915c, 7).

[13]This paper (Riecke 1915) appeared with an adulatory introduction by Debye, which in the original reads: "Mit stets steigendem Interesse und jugendlicher Frische verfolgte Riecke buchstäblich bis zu seinen letzten Tagen die neuen weittragenden Errungenschaften, welche der physikalischen Forschung durch die Erweiterung der Einflußsphäre des Planckschen Wirkungsquantums beschieden waren. Das viele Unklare und Unbefriedigende, das noch unüberwunden sich in diesem Gebiete einem restlosen Verstehen entgegenstellte, war ihm duchaus nicht ein Grund zur überkritischen Ablehnung. Vielmehr schöpfte er gerade daraus einen wesentlichen Teil der Freude, welche

were the strongest players, and both had build up strategic relations to neighboring fields. Klein was an eager advocate of linking mathematics to technical applications, while Hilbert focused on theoretical physics, which he wanted to make more rigorous with the help of mathematics. Hilbert is often considered the preeminent mathematician of the twentieth century, who founded influential and fruitful research programs that continue to enthrall mathematicians up to the present day. Like Voigt, his senior by twelve years, the only two sites of activity in Hilbert's life were Königsberg and Göttingen. In Königsberg he had to wait six years as a *Privatdozent* before becoming an associate professor, finally receiving a full professorship in 1893. "In 1895, 33 years old, I was called to Göttingen by Felix Klein," he once wrote, "essentially, I think, due to my work on the theory of invariants; the central problems of it I had tackled and solved from a novel point of view."[14] Hilbert gained great prominence with his axiomatization of Euclidean geometry in 1899, and even more in 1900, when he gave a speech in Paris listing problems that set an agenda for generations of mathematicians. Making his point pronouncedly against Emil du Bois-Reymond, he declared that there is no ignorabimus, at least not in mathematics. While Hilbert's vigorous scientific leadership is undisputed, at the same time he is often seen as an rather quixotic and even childish genius.[15] With his spectacular successes at the turn of the century, Hilbert monopolized Göttingen's reputation for mathematical progress. Klein stepped aside, concentrating on his role as a science organizer.[16]

As it may not yet be clear why Hilbert should be of importance for a study on the early history of quantum theory in Göttingen, it should be obvious that incorporating the mathematical dimension of this development is necessary. Many later quantum physicists started as mathematicians, and leading mathematical figures like Hilbert and Henri Poincaré included physics in their greater mathematical visions—the axiomatization of physics was simply Hilbert's "sixth problem."

---

ihm seine Beschäftigung mit dem Atominneren eingebracht hat.

Restlos überzeugt von der großen Wichtigkeit und Tragweite der neuen Ideen, war es ihm unmöglich als müßiger Zuschauer abseits am Wege zu stehen. Er empfand das Bedürfnis zunächst in möglichst weiten Kreisen das Interesse für die reizvollen neuen Probleme, welche hier zum Greifen nahe liegen, zu beleben und zu wecken. In dieser Absicht und in diesem Sinne verfaßte er in der allerletzten Zeit vor der Krankheit, die mit dem Tode endigen sollte, die folgende Notiz, welche in knapper Form das Wesentliche der neueren Ansichten über das Entstehen und die Gesetzmäßigkeiten der Spektren darstellt."

[14]Document 19 (p. 105.)

[15]For a biography of Hilbert cf. Reid (1970). Note, however, that this (as acknowledged in the foreword of the second edition) is a "romantic" book that lives from a certain "mathematical innocence" at the time of writing. Nonetheless, it has benefited much from the assistance of Richard Courant and Paul Ewald and is a quite reliable source of information about Hilbert's life, despite the romanticization of his personality. Unfortunately, no references are given. Some attempts to shed light on Hilbert's personality by dismissing the traditional and unverified pictures of a naive eccentric or ivory tower dreamer can be found in Rowe (1992).

[16]Rowe's distinction between Hilbert as the pure *Fachmathematiker* (a mathematician only concerned with his field) and Klein as the *Wissenschaftspolitiker* (science politician) in Rowe (1989, 201f), will be scrutinized in the following.

For a first comparison of Voigt and Hilbert two points shall suffice. Their different administrative engagement in the university might strike one's eye first, as Voigt repeatedly held official posts like that of the rector of the university or dean of the faculty, while Hilbert was strikingly abstinent in this respect. When it came to important decisions, however, this was not the case. In his autobiographical sketch he wrote:

> In all respects of organization Klein undisputedly had the leadership; I never cared about administrative things. But when it came to essential decisions, in particular on appointments, on the creation of new positions and the same, I always took an active interest.[17]

Secondly, there were the frequent offers of appointments to Hilbert that gave him weight in negotiations with the ministry about improvements to his conditions in Göttingen, critical weight Voigt badly needed.[18]

Hilbert's relationships with the other Göttingen professors was not always smooth, and even towards Klein Hilbert himself did not gloss over the "differences in temperament."[19] In a dispute about the new professor Friedrich Schilling in 1900, Hilbert fought against Klein and Voigt. While the former "gave up his hangover idea [Kateridee] in time on his own and—as must be acknowledged—unconditionally and for all times" the latter "who with pleasure and on all occasions inveighs against Klein behind his back but fails completely when it is necessary to speak up," remained opposed, to Hilbert's annoyance.[20] Without question, the often invoked picture of a Göttingen community of mathematicians and physicists of mutual interests and forces is a untenable hypothesis.

As a consequence of this particular assemblage of resources, interests, personalities and their coalitions in Göttingen, which is already apparent from the few examples mentioned, one must expect that the history of Göttingen's ascent to a center of quantum physics during the first quarter of the twentieth century will turn out to be the result of a rather complex interaction of forces that steered the whole course in one direction, not fully foreseen by the different players, who pushed in their various directions.

---

[17]Document 19 (p. 105).

[18]Hilbert lists calls from Berlin (multiple), Leipzig, Heidelberg, and Bern, cf. document 19.

[19]Document 19 (p. 105).

[20]The strongly worded original reads: "Die Affaire Schilling ist zu aller Zufriedenheit erledigt. Schilling bleibt—aber als Extraordinarius. Für mich hat die Sache das Gute gehabt, dass ich die Gewissheit erhalten habe, dass ich mich in solchen Fragen auf die Fakultät absolut verlassen kann; sie hätte wie ein Mann auf meiner Seite gestanden—trotz Voigt, der ja stets mit Behagen und bei jeder Gelegenheit hinter Klein's Rücken auf diesen raisoniert, aber, wenn es gilt ihm entgegenzutreten, völlig versagt. Voigt habe ich meine Meinung gesagt. Klein gab selbständig rechtzeitig und—wie anerkannt werden muss—bedingungslos und für alle Zeit seine Kateridee auf; es ist ihm, glaube ich, überhaupt unangenehm, dass er sie je gehabt hat." Hilbert to Sommerfeld, 27 September 1900, DMA HS 1977–28/A, 141.

In order to exhibit this parallelogram of forces, which at the same time relates to a kind of subtle economy of resources, I will proceed in four steps. First, I analyze the opportunities offered by retirements and new positions for a new alignment of research areas and personnel in the years before the First World War. Then I try to identify a particular "vision" that was at the heart of this reorganization of Göttingen physics and attribute it to Hilbert. Here I also reconstruct the role of Debye, who for the first time tried through research and teaching to establish quantum theory in Göttingen even during the war. Next, I look at the research groups or "schools" of Born in Berlin, Frankfurt and Göttingen, in which the experiment played are far greater role that previously recognized. And finally, I suggest that after 1920 it was not the one golden road that led to a singular revolutionary achievement, but five avenues which, although they did not meet exactly in one point, still brought about matrix mechanics and, beyond that, the statistical interpretation and more. In this way, I try to explain to what extent the new quantum mechanics was the result of long-term groundbreaking work also in institutional terms.

Chapter 2
From Generational Change to Scientific Opportunity

Scientific change is most often also a change of scientists, but does the inverse also make sense? Perhaps this may be the case if the changes are made intentionally. Thus the first field of research politics—which does not necessarily add up to a stringent policy—concerns the discussions about, and even fights for candidates and successors at the Göttingen faculty. In the following, I begin by describing the changes in personnel of the chairs of physics and neighboring fields. These are included to illustrate the rather exceptional changes in some fields like physics. The analysis of continuities and discontinuities is then employed to identify the driving forces behind the observed changes.

2.1 The Dynamics of the Main Mathematics and Physics Positions, 1900–1926

German ministerial and university bureaucracy typically have kept detailed records about personnel matters, as the formal processes for filling professorial positions were complex.[1] Therefore there is ample material to reconstruct the development of the staff in the physical and mathematical sciences, at least according to the "official" reading. I will proceed in describing the institutional and personnel dynamics in this way, starting with mathematics and then turning to theoretical, experimental and applied physics as well as related fields, in order to finally arrive at a pattern of structural change. This will in particular exhibit some crucial instances where local conditions and specific actors have had an influence.

---

[1]For this chapter I used the following archival sources in particular: Geheimes Staatsarchiv Preußischer Kulturbesitz Berlin (=GStA PK), Rep. 76 V a, Sekt. 6, Tit. IV, Nr. 1, Bd. XXIV–XXVIII (correspondence with the ministry of culture); Universitätsarchiv Göttingen (=UA-Gö), Kur. 4 I 105 (formerly XVI.V.B.7), Vol. I–II (correspondence with the university curator); Kur. 4 Vc and Kur. P.A. (staff files); Phil. Fak. II Ph. 36 a–f (faculty issues); Niedersächsische Staats- und Universitätsbibliothek, Special Collections, Cod. Ms. D. Hilbert, F. Klein, W. Voigt, K. Schwarzschild (= Hilbert Papers, etc.).

© The Author(s), under exclusive license to Springer Nature Switzerland AG 2019
A. Schirrmacher, *Establishing Quantum Physics in Göttingen*,
SpringerBriefs in History of Science and Technology,
https://doi.org/10.1007/978-3-030-22727-2_2

Mathematics

From 1904 the mathematics section of the philosophical faculty was represented with four full professorships, relegating Berlin to second place in mathematics (Lorey 1916; Rowe 1992, 1985, 436). The extent to which Göttingen dominated the field of mathematics is apparent, for example, in Ferdinand Frobenius' request to the Ministry of Culture to finance a fourth chair in Berlin and to appoint Hilbert in 1914. Frobenius wrote that while Hilbert had declined the offer in 1902 "in the meantime, however, after Berlin has become a little Göttingen itself, he might have changed his mind."[2] Hilbert remained in Gauss's chair until 1930 when Hermann Weyl became his successor, not really a close disciple but near to his understanding of mathematics.

The next important mathematics position in Göttingen was that of Felix Klein, who had to retire early in 1913 for health reasons. Constantin Carathéodory was his direct successor, then Erwin Hecke took over the position in 1917, followed by Richard Courant in 1920. Weyl and Brouwer were also present on all the three lists of candidates prepared by the committees for the ministry. Gustav Herglotz was also suggested in 1919, but eventually none of the three actually came. After Weyl took half a year to consider the offer and finally decided against it, a further proposal became necessary. This enabled Courant to return to Göttingen. Interestingly, almost all of the proposed candidates were students of Hilbert (only Carathéodory did his doctorate with Minkowski), and none of them of Klein.

In 1902 Hilbert had succeeded in trading a call to the prestigious Berlin university for a new full professorship for Hermann Minkowski. After his sudden death, Minkowski was succeeded by Edmund Landau in 1909. Though Hilbert was eager to find a position for Zermelo, Landau was his choice, as he was probably Klein's.[3] And the fourth position the one created for Carl Runge in the new field of applied mathematics, this time at the instigation of Klein in 1904, who was at excellent terms with Friedrich Althoff, the most influential planner of faculty hiring at Prussian universities. Runge also played a certain role for theoretical physics, as he was interested in spectra.[4] And in 1925 Gustav Herglotz, although a representative of pure mathematics, became Runge's successor in applied mathematics, augmented by astronomy.

---

[2]Frobenius to Ministry 22 June 1914, reprinted in Biermann (1988, 325): "Damals [1902] hat er den Ruf abgelehnt; inzwischen hat er sich, nachdem Berlin ein kleines Göttingen geworden ist, vielleicht eines besseren besonnen."

[3]The faculty named Adolf Hurwitz, Otto Blumenthal and Edmund Landau in no particular order, cf. Rowe (1992, 564 f.); it has been suggested that Hilbert supported Landau, cf. Peckhaus (1990, 121). In an undated letter to Klein in 1909, Hilbert neutrally writes that Landau accepted the call and immediately afterwards reports that he was successful in increasing Zermelo's remuneration considerably. He adds that Landau will lecture mainly on prime numbers, "Crooked lines and surfaces do not suit him at the moment" (Frei 1985, 138). In a letter to Sommerfeld, however, Hilbert stated that Landau was his choice: "Fakultät und Ministerium legten die Wahl ganz in meine Hände; ich hoffe, dass sie gut ausgefallen ist." Hilbert to Sommerfeld, 10 April 1909, (Sommerfeld 2000, 356–358, on 357).

[4]Cp. the article (Runge 1907) which was presented in the Göttingen Mathematische Gesellschaft on 7 May 1907.

Physics

In contrast to the mathematicians, the years from 1914 to 1921 saw much greater change in the main physics positions. The changes resulted in rather complicated patterns of replacements and successions. This dynamics in personnel resources reflects difficulties in acquiring proper successors and, probably, also differences about prospective research programs as well as opportunities for exerting influence to set a new course.

The two leading physicists in Göttingen, Woldemar Voigt and Eduard Riecke, set great store on smooth cooperation, as neither of them restricted himself to the experimental or theoretical domain alone, attending to both lecturing duties and research projects. Yet formally, Riecke had the experimental chair and Voigt the theoretical one. Both had become ordinary professors of physics in Göttingen around 1880. Riecke was mostly interested in the corpuscular structure of electricity, Geissler tubes and the theory of conductivity, while Voigt, who was probably the most productive theoretical physicist in the 1880s, worked on the properties of crystals and their interaction with light (Jungnickel and McCormmach 1986, 112–124).

Riecke and Voigt had intended to retire at the same time in 1915 to make room for first-class replacements. Peter Debye, however, was called to a position that had not existed before, but was created for him in 1914. The idea was that he might take Voigt's chair the following year, but due to the war, Voigt had to teach up to his death in 1919, as had Riecke, who died in 1915. The circumstances of Debye's appointment will be detailed in Sect. 3.5 against the larger backdrop of the general developments in Göttingen in both mathematics and physics.

For Riecke's succession four physicists were nominated, all of whom coincidentally were associated with research on quantum problems: Wilhelm Wien in first place, then Friedrich Paschen, and Johannes Stark and Pieter Zeeman equally on the third rank. However, none of these men came. Wien was looking ahead to move on to Röntgen's prestigious chair in Munich, Paschen negotiated too long, such that the ministry finally rescinded the offer, while Stark was not welcome to one wing of the faculty as Zeeman was not to the other.[5]

The fact that Voigt's chair did not become available as planned due to the war, and the insight that Debye's experimentalist capabilities qualified him beyond theory, let the move appear attractive to make him Riecke's successor instead. This in turn meant that another subordinate position opened for the teaching of students from other fields like medicine and chemistry. For this position, younger candidates such as Oskar Emil Meyer were nominated first, and Robert Pohl and Peter Paul Koch second. After considering an even larger circle, James Franck, Egon Schweidler, Hans Geiger and Eduard Grüneisen, Pohl was finally chosen in 1916, but only took up his position after the war and was promoted to full professorship thereafter in 1920 (Table 2.1).

In the same year, Max Born (with Erwin Madelung and Wilhelm Lenz as alternatives) was appointed successor of Debye, who was lured away by the rivaling ETH Zurich and Switzerland's better living conditions directly after World War I. Initially,

---

[5]See next section below with more details.

Table 2.1 Candidates for mathematics and physics professorships 1900–1926

Position	Candidates	Notes
New position 1902	**Minkowski**, Königsberg	Hilbert's condition to stay
Minkowski 1909	Hurwitz, Zurich Blumenthal, Aachen **Landau**, Berlin	All placed on equal level, all of Jewish decent
Schwarzschild 1909	1. **Hartmann**, Potsdam 2. Cohn, Königsberg 3. Kapteyn, Groningen	
Klein 1913	1. **Carathéodory**, Breslau 2. Weyl, Göttingen and Brouwer, Amsterdam	
No free position 1914 (Voigt 1915)	**Debye**	Plan: early replacement for Voigt, not Riecke
Riecke 1914	1. W. Wien, Würzburg 2. Paschen, Tübingen 3. "first" Stark, Aachen, "second" Zeeman—Amsterdam	Simon, von Seelhorst, G. Frölich, G. E. Müller, Lorenz, Mossbach, Pohlenz, I. Hartmann against placing foreigner Zeeman
– New list 1915	1. E. Meyer, Tübingen 2. **Pohl**, Berlin, P. P. Koch, Munich [Franck]* [Schweidler, Geiger, Grüneisen]**	
Wallach 1915	1. Willstätter, KWI "unavailable" (declined 23 June) 2. Thiele, Strasbourg (28 June) 3. Harries, Kiel (8 July); Auwers, Greifswald; Knorr, Jena	
– New list	1. **Windaus**, Innsbruck 2. Wieland, Munich 3. Diels, Berlin 4. Braun, Breslau ...	
Caratheodry 1917	1. **Hecke**, Basel 2. Brouwer, Amsterdam and Weyl- Zurich 3. Blaschke, Königsberg	
Hecke 1919	1. Brouwer, Amsterdam 2. Herglotz, Leipzig 3. Weyl, Zurich	Statement of Hilbert
– Added 1920	4. **Courant**, Münster	
Simon 1919	1. M. Wien, Jena 2. Gaede, Freiburg	The prestigious chair for technology created by Klein now given to a local *Privatdozent*
– New list	1. Prof. Dr.-Ing. Rüdenberg, Oberingenieur at Siemens-Schuckert and PD at TH Berlin 2. Barkhausen, Dresden 3. Möller PD Hamburg or **Reich**, PD Göttingen	
Debye 1920 (Voigt 1915)	1. Sommerfeld, Munich (2. Born, Frankfurt 3. Mie, Halle)	Officially Riecke's old position for experimental physics, Born argued that Voigt's position had yet to be filled
– Modified list	1. **Born**, Frankfurt 2. Madelung, Kiel 3. Lenz, München + **Franck**, Berlin	
Hartman 1921/1927	**Kienle**, Göttingen Wirtz, Kiel Graff, Hamburg	
Runge 1925	**Herglotz**, Leipzig	

*Mentioned in interview Debye 1962, p. II/5 (AHQP)

**Invited for talks in winter term 1915 of candidates for the position, details see note 90 Chapter 3 (p. 53)

the Göttingen faculty had suggested the more senior physicists Sommerfeld and Mie alongside Born, but after the first choice Sommerfeld declined, the list was rewritten. Born succeeded in getting James Franck called for an additional full professorship, arguing that Voigt's position had not been filled after his death (although it had been traded for the creation of new and enhancing existing positions).[6]

Additional physics positions were Emil Wiechert's in geophysics and Karl Schwarzschild's in astronomy. Wiechert came to Göttingen in 1897 and soon became extraordinary professor of geophysics, though his more important work was in electron theory and he was known to move in the direction of pure physics. He was promoted to full professor in 1904, at a time when he was offered the Munich chair of theoretical physics. With his promotion in Göttingen, he held the chair of geophysics until 1928. Schwarzschild, who came as director of the Göttingen observatory in 1901 and soon became full professor, contributed to questions of general physics, too, but unlike Wiechert, he was lost to the astrophysical observatory in Potsdam early in 1909. He was succeeded by Johannes Hartmann, yet his activities flagged through the war and he eventually left his position in 1921. A successor, Hans Kienle, was not appointed until 1927.

Applied Sciences

While specialized fields like geophysics and astronomy were pursued by their Göttingen proponents Wiechert and Schwarzschild with fundamental physics in mind, further positions in applied fields had been created on the initiative of Felix Klein, in particular. While the position for technical physics, comprising fields like agricultural machines and aeronautics, was Ludwig Prandtl's from 1904 for over 40 years, the one for applied electricity was held until 1919 by Hermann Simon. Particular problems arose in finding an appropriate replacement for Simon. After the failure of the proposal to hire Max Wien and Wolfgang Gaede, a second list was submitted with Reinhold Rüdenberg, a Siemens engineer, on the top position, then Heinrich Barkhausen and finally Hans Georg Möller, who was preferred even over Max Reich, the last choice. It was Reich, however, Simon's former assistant, a Göttingen *Privatdozent*, and only of second-rank quality, who would become representative of the field of advanced technical applications once so promoted by Klein in the early Weimar period.

Chemistry

Besides applications in technology, chemistry was the other great field that was to develop ever closer relations to physics and its new theories on atoms and quanta. At that time, Chemistry in Göttingen was in the process of becoming a club of Nobel laureates. Physical chemistry was introduced by Walther Nernst, winner in 1920, who, however, left for Berlin in 1905. His successor was first Friedrich Dolezalek, then Richard Zsigmondy in 1907, who received the Nobel Prize in 1925. After futile efforts to win Clemens Winkler, William Ramsey (NP 1904) and Theodore

---

[6]Slightly differing accounts can be found in Hund (1987), Jungnickel and McCormmach (1986), Dahms (2002), Greenspan (2005).

W. Richards (NP 1914) for a chair of inorganic chemistry, Gustav Tammann was hired in 1903, who left this field to Zsygmondy and instead dealt with physical chemistry. Almost as difficult as the Simon succession, but with a better outcome, was the case for Otto Wallach's organic chemistry chair. In 1915, the Nobel Prize winner of 1910 was replaced by the later laureate of 1928, Adolf Windaus.

Scientific Philosophy

The parallel developments of philosophy positions that were related to mathematics and the natural sciences show similarities to the problems in the field of physics, also with regard to the involvement by mathematicians. When the Prussian Ministry suggested to the Göttingen philosophical faculty a third professorship for Edmund Husserl in 1900, this met with stiff opposition from the philosophers—who were largely in favor of experimental psychology and philology. A year later, however, they were ignored when Husserl, a philosopher trained in mathematics and interested in foundational issues, became an extraordinary professor.[7] He entered into a fruitful exchange with Hilbert in the *Mathematische Gesellschaft*. When the third *Ordinariat* was proposed again in 1908, Paul Natorp, Heinrich Maier, and Ernst Cassirer were put on the list of the faculty. Hilbert demanded, however, the promotion of Husserl in this position, thus causing persistent trouble in the faculty. Maier was called only in 1911 and left for Berlin in 1918, which again caused major problems between the "historical-hermeneutical" and the "natural-scientific" wings of the faculty: Husserl accepted an offer from Freiburg for a full professorship in 1916. Hilbert tried hard to secure this position for Leonard Nelson, also a mathematically trained philosopher. Finally, Georg Misch was called to Husserl's position in 1918 and promoted to Maier's in 1919, making room to confer an extraordinary professorship on Nelson at the same time. The list of natural scientists named Moritz Schlick second. Courant even suggested that Hermann Weyl succeed Maier,[8] as Hilbert had nominated the vitalist philosopher Hans Driesch.[9] As with the physics chairs, the part of the philosophy faculty that dealt with science and mathematics also underwent considerable change. This fact suggests that mathematicians and physicists of the philosophical faculty have tried to establish a kind of philosophy to their taste.

And a Host of Privatdozenten

This picture of the human resources of Göttingen physics and neighboring fields would not be nearly appropriate without mentioning the relevant staff in Göttingen below the professorial rank. Typically, the aspirants for future professorships play a decisive role for the scientific potential of a research location. In Göttingen the large number of *Privatdozenten* turned out to be a special characteristic. As David Rowe

---

[7]Cf. also for the following: (Peckhaus 1990, 208ff).

[8]Courant to Hilbert, July 1918, as cited in Reid (1970, 90).

[9]Cf. the answer Becker to Hilbert, 10 October 1918, UA-Gö, Hilbert Papers, folder 15A, Bl. 1/1–1/2.: "Ihre Wünsche gehen aber darin zu weit, wenn Sie meinen, mehr oder weniger sämtliche Lehrstühle der Philosophie in Göttingen mit naturwissenschaftlich orientierten Philosophen besetzen zu können."

Table 2.2 *Privatdozenten* in Göttingen 1905–1925. Years in italics indicate "außerplanmäßige (nichtbeamtete außerordentliche)" professorships, i.e. special non-permanent lecturing positions without many professorial privileges; underlined years indicate permanent associate professorships. Data from *Catalogus professorum gottingensium* 1734–1962, W. Ebel, ed., Göttingen 1962, and *Amtliches Verzeichnis des Personals und der Studierenden der Kgl. Georg-Augusts-Universität zu Göttingen*, Göttingen 1905–1917

Mathematics	Mathematical/ theor. physics	Applied math. technical physics	Experimental physics	Physical chemistry
Carathéodory 04–08 Koebe 07–10	Abraham 1900–09	Herglotz 04–07	Stark 1900–06 Krüger 06–09	Bose 01–06
Bernstein 07–11 Toeplitz 07–13 H. Müller 08–10 Weyl 10–13 Schimmack 11–12 (didactics)	Ritz 09 Born 09–14 Haar 10–11 (Hellinger 07–09 nicht PD)	v. Kármán 10–13	Bestelmeyer 06–17 Gerdien 07–30 (at Siemens since 1908) Rümelin 11–14	Coehn 1899–1919, *21–28* scientific philosophy
Hecke 12–15 Courant 12–19 Behrens 13–17	Madelung 12–19 P. Hertz 12–21, *21–33*	Reich 12–20 v. Sanden 11–18	Rausch Traubenberg 12–21, *21–22* L. Geiger 12–19 Gerlach 17–18	v. Nelson 09–19, *21–27*
Bernays 19–22 *22–33* Noether 19–22, *22–33* Neder 20–22 Behmann 21–25 Siegel 21–22	H. Kneser 22–25	König 20–22 Betz 22–26, *26–?* Nádai 23–26, *26–29*	Scherrer 19–20 Busch 20–21 Grotrian 21–24 Gudden 21–24, *24–26* Oldenberg 22–26, *26–33*	Lipps 22–28, *28–36*
Ostrowski 23–27	Heisenberg 24–*26*	Schuler 24–28, *28–41* Walther 24–28	Goetz 23–29, *29–39*	

has put it, Göttingen's wealth in qualified candidates in the mathematical sciences was striking:

> Most impressive of all, between 1890 and 1914 Göttingen had no fewer than eighteen *Privatdozenten* whose names read like a "Who's Who in German Science" during the Weimar era [...]. This is in contrast with the situation in Berlin, where between 1897 and 1901 there were no *Privatdozenten* whatsoever (Rowe 1989, 202).

The same applied to the later years. The picture that emerges for the years of nascent interest in quantum theory up to the establishment of quantum mechanics is given in Table 2.2. In this table, mathematical subfields are not distinguished; in physics, the distinction is made between the three divisions of the physics department, which were headed by Max Born, James Franck and Robert Pohl in the 1920s.

A special quality of Göttingen in the field of mathematical and physical sciences was hence that besides the strong group of professors *in office*, there was an equally large and strong group of professors *in spe*. They had the right (and duty) to teach and would do so mainly in the particular fields of their research interests. Apparently also in Göttingen self-recruitment was widespread, because sooner or later scholars like Theodore Caratheodory, Hermann Weyl, Erich Hecke and Richard Courant from the mathematicians and the physicists Max Born, Max Reich and Werner Heisenberg (however, only after 1945 and for very special reasons) would return to the university as full professors where they had earned the qualification for this position.[10] Some of the *Privatdozenten* were also able to become special extraordinary or rather *außer-planmäßige* professors, though only outside the ordinary positions and without the pay and privileges of a civil servant, if they did not find such a post elsewhere and the Göttingen colleagues succeeded in financing it through special regulations.

In this way a number of scholars became professors in the crisis years after the Great War and the Revolution. Paul Bernays, who wrote a second habilitation thesis on axiomatics in 1922 (after he had already done so in Zurich with a work on function theory) became extraordinary professor of mathematics without tenure in the same year, as did Emmy Noether, who, as a woman, had to fight for many years to be granted the habilitation at all. The physicist Paul Hertz acquired the title of professor a year before; however, his field had shifted into a more philosophical direction, while Leonard Nelson, a philosopher who did his habilitation with David Hilbert back in 1909, but never found a position in philosophy until he was given an extraordinary professorship in the mathematics section of the philosophical faculty in 1919, became tenured in 1921.[11]

2.2   From Structures to Actors

The overview of positions and staff of physics, mathematics and related fields in Göttingen has provided a particularly rich ensemble of scientists, many of whom were able and sometimes active in various research fields outside the formal description of their appointments. Moreover, experimental and theoretical physics, often characterized as pure physics or physics proper, and later famously represented by the triple professors Born, Franck and Pohl, was indeed widely framed by a host of specialized and applied fields as well as related mathematical and philosophical projects. I will therefore identify a striking pattern or structure in the dynamics of this academic field, which will allow us to discover breaks, discontinuities and trans-formations, which then shall be analyzed more thoroughly from a perspective of the relevant actors.

---

[10] Note that the German system is a strictly non-tenure track, so that a scholar who did his habilitation at one university first needs to get a position at another in order to return to a professorship.

[11] See Ebel (1962) and Tollmien (1991).

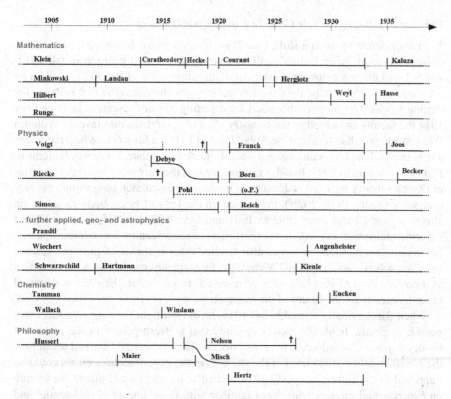

Fig. 2.1 Dynamics of the main mathematics and physics positions as well as from related fields in chemistry and philosophy 1905–1935 (broken lines indicate extensions beyond regular duty or precursory appointments to ordinary positions; † deceased)

If one summarizes the information collected on the dynamics of research positions in physics and neighboring fields at Göttingen between the turn of the century, and at the time when quantum mechanics was established in 1925 and 1926, in a pictorial way, one can easily observe how new lines emerge while others are discontinued (Fig. 2.1). This indicates dates when decisions for changes in the focus of research fields occurred. On the one hand, one observes a fairly constant background configuration of positions in mathematics, chemistry, and some physics like geophysics and technical physics. On the other hand, significant problems in personnel politics can be read from the graphical representation in the core physics positions: except for Wiechert's geophysics and Prandtl's technical physics, no other physics position preserved a continuity in the research field covered under the developments from 1914 to 1921. The broken lines in this picture, roughly speaking, represent the possibilities to establish new research programs; they may be considered in this sense as necessary, but clearly not as sufficient conditions for a shift of focus in the research interests of the physics faculty.

The Fate of Eduard Riecke's Chair in Experimental Physics

To give evidence for such a shift, I now turn to some more details on a particularly difficult position. After Debye's call to a newly established position in fall 1914, which I will discuss below in greater detail, the succession of Riecke, the holder of what was understood to be the experimental physics chair, was the first instance for turning towards physicists who stood for quantum physics research. In December 1914 the faculty presented to the ministry a list of candidates that favored Wilhelm Wien. Wien, who had received the Nobel Prize in 1911 for his work on heat radiation, a field in which he had been engaged since 1893, tried to relate X-ray wavelengths to Planck's quantum in 1907. Besides Lorentz, he was the only other weighty supporter of Planck's theory by 1910 and had been feeding the discussion on quantum physics ever since (Kuhn 1978, 202ff). Publishing on the laws of black-body radiation on almost a yearly basis from 1893 to 1901 and from 1907 to 1915, Wien exhibited experimental as well as theoretical expertise that became central for quantum matters. For certain, his publication record also covered other branches of physics—among them sea waves, canal rays, and X-rays, yet by no means can his work be identified as a continuation of Riecke's lines of research. In particular, their views departed most drastically on the future of the quantum.

When Riecke finally retired in fall 1914, he did not fight for a specific successor nor for a specific field, but merely required that a "fresh person in the middle of his development" should take over his position. He was more concerned about how the general mathematical level of the students in the experimental lectures could be improved in the future, suggesting "in particular no one should attend the lecture on experimental physics who is not familiar with the elements of differential and integral calculus."[12]

It appears that Riecke's chair was never officially offered to Wien.[13] As Hilbert pointed out to the ministry early in 1915, in his view Wien was the only one still missing in the new Göttingen physics team, at the same time, he did not believe that his appointment would be probable during the war.[14] To the ministry it may have seemed futile to negotiate with Wien, who was Röntgen's presumptive successor. One year later Wien mentioned that he actually might have considered such an offer independently of his chances in Munich. But he conceded: "I cannot say, however, whether negotiations with me would have succeeded, since the conditions at the Göttingen institute are in fact not very favorable. The premises of experimental physics are relatively small since the theoretical physics has occupied as much."[15]

As early as one year in advance, the simultaneous retirement of Riecke and Voigt had been agreed upon, and since that time the first priority had been to find a single

---

[12]Riecke to Ministry, 4 October 1914, requests that a "frische, noch mitten in der entwicklung stehende kraft an meine stelle tritt;" complains about "sehr ungleiche vorbereitung der zuhörer." GStA PK Rep. 76 V a, Sekt. 6, Tit. IV, Nr. 1, Bd. XXIV, Bl. 108–109v.

[13]Ministry to Göttingen Curator, 2 July 1915, writes that appointing Wien was "hopeless." GStA PK Rep. 76 V a, Sekt. 6, Tit. IV, Nr. 1, Bd. XXIV, Bl. 181.

[14]Document 10 (p. 101).

[15]Document 14 (p. 103).

new director of outstanding quality.[16] When it seemed clear that Wien was out of reach, Paschen became the alternative.

In particular, Paschen met the requirement to be able to combine experiment and theory in a way Voigt and Riecke had performed collectively; for them the understanding of theoretical physics went back to Franz (and Carl) Neumann, who did not find it in spectacular hypotheses but rather careful sifting existing evidence and bringing it into mathematical order.[17] A new kind of collaboration that differed considerably from the Neumann model had emerged between Paschen and Sommerfeld in late 1915.[18] This was, however, only after the offer of the Riecke chair and long negotiations had finally failed.

Paschen's research in his early career at Hanover stood in competition with the Berlin group, which was working on black-body radiation, viz. Heinrich Rubens, Otto Lummer, Ernst Pringsheim, Max Thiesen, Eugen Jahnke and in particular Wilhelm Wien). As Paschen derived the same law as the latter and communicated it to him before its publication, it might well have been called the Paschen-Wien law.[19]

Paschen was in several respects a Berlin-independent physicist dealing with quanta and spectra. As Paul Forman has pointed out: "In striking and curious contrast with the Berlin experimentalists, who were literally enraged at him, throughout his work on the black-body radiation problem, Paschen the pure experimentalist showed himself to be more than ready to enlist experiment in the service of theory" (Forman 1970b, 346). In addition, he collaborated with Runge on spectroscopy. After his appointment in Tübingen and a phase of work on radioactivity, X-rays, canal rays, and the mechanism of light emission, he returned to spectral series in 1908. Motivated by Göttingen student Walther Ritz, who spent the winter term 1907/08 in Tübingen, he looked for and found what later became known as the "Paschen series." The Paschen-Back effect of 1912, again, was based on Ritz conceptions, and "was immediately seized upon as one of the potentially most revealing clues to atomic structure and the mechanism of emission of spectral lines" (Forman 1970b, 346). In 1914 Paschen began work on "Bohr's helium lines," an occupation that was to distract him from hurrying to the colors. It had become of utmost importance to him when Sommerfeld was inquiring about data for his theory, thus starting a fruitful collaboration in November 1915.[20]

---

[16]Curator to Minister, 18 April 1914, GStA PK Rep. 76 V a, Sekt. 6, Tit. IV, Nr. 1, Bd. XXIV, Bl. 69–70.

[17]See on Voigt's teacher Franz Neumann and also on his son Carl Neumann, a professor of mathematical physics in Leipzig, (Jungnickel and McCormmach 1986).

[18]Cf. Paschen to Sommerfeld, 14 November 190, DMA HS 1977–28/A,253, who reflects upon the relationship between the theoretician and the experimentalist, questioning whether any pure experimentalist exists ("reinen Beobachtungskünstler"), since even experimental physicists can do experiments in a theoretical manner ("theoretisch experimentieren").

[19]Cf. Kangro (1969), compare also Heinrich Kayser's critical view on Wien's work in his *Erinnerungen*, (Kayser 1996, 136): "Through Wien's radiation law' he became a great man and he was supported by the fact that about at the same time Paschen took great pains over deducing the same law experimentally."

[20]Forman (1970b, 346), Paschen-Sommerfeld correspondence (Sommerfeld 2000; Paschen 1916).

Paschen had considered the Göttingen offer seriously and it had turned out in the negotiations that securing working conditions in spectroscopy was a central point. But like Wien, he saw resources inequitably distributed between the experimental and theoretical institutes. For example, the new diffraction grating bought with support of the *Göttinger Vereinigung* promoting applied physics and mathematics in 1911 came to Voigt's facilities, and was presented to the public in a paper on which he collaborated with a mathematician (Voigt and Hansen 1912). Under the condition of a new allocation of rooms and that "instruments for spectroscopy" would be transferred to him, he signed an agreement with the Prussian ministry to accept the Göttingen offer. Using a four-day respite, he negotiated with the Württemberg ministry after his conditions in Göttingen were met and then turned again to Berlin to ask for post-war budget guarantees. As a result, according to Ludwig Elster, the official in charge, Paschen "couldn't make up his mind" and the ministry asked for a new list of candidates, making Hilbert and Debye hurry to Berlin.[21] Having overplayed his hand, Paschen stayed in Tübingen with a lower salary than Göttingen had offered.[22] As a consequence, one may argue on the basis of the greatly increased correspondence with Arnold Sommerfeld on atomic spectra and models, Paschen intensified his collaboration with Munich physics instead until his promotion to president of the Berlin *Physikalisch-Technische Reichsanstalt* in 1924 (Sommerfeld 2000, 469ff.).

On their first list the members of the Göttingen commission had indicated that concerning Wien and Paschen they would "value the abilities of the above character-ized men so highly, [...] that we would rather restrict ourselves to mentioning their names." In the third position they then added Stark "first" and Zeeman "second." They explained, "We can, however, not suppress certain doubts in other respects [than their scientific abilities] against them."[23]

Immediately after Paschen's withdrawal, Stark claimed that it was now his turn. His many letters to the ministry, in which he alternated between pointing out his ability to foster "German physics in its current crisis," the humid weather in Aachen that was harmful to his wife's health, and the allegation by Göttingen staff that he was an anti-Semite, admitting, "of course I fight against one-sided philo-Semitic attempts in academic circles."[24]

After things remained unresolved under these circumstances, Wien finally stepped in by writing to Otto Naumann at the Ministry in December 1915. He tried to take the initiative in order to solve the problem of the full professorship in experimental physics in Göttingen, as he saw it, "in the general interest of our science." This

---

[21] Agreement, Paschen with Elster, 19 June 1915. Paschen to Elster and Debye to Elster, 23 June 1915. Paschen to Elster, 27 June 1915. Elster to Curator and Elster to Voigt 2 July 1915. Telegram Hilbert and Debye to Elster, 6 July 1915. GSPtKB Rep. 76 V a, Sekt. 6, Tit. IV, Nr. 1, Bd. XXIV, Bl. 307–338.

[22] Göttingen offered 8400 M basic salary, Tübingen 6500M. Agreement Elster with Paschen, 19 June 1915, GStA PK Rep. 76 V a, Sekt. 6, Tit. IV, Nr. 1, Bd. XXIV, Bl. 307–309; (Forman 1970b).

[23] Dean to Minister, 24 December 1914, GStA PK Rep. 76 V a, Sekt. 6, Tit. IV, Nr. 1, Bd. XXIV, Bl. 124–126v.

[24] Stark to Ministry, 18 June and 1 July 1915, GStA PK Rep. 76 V a, Sekt. 6, Tit. IV, Nr. 1, Bd. XXIV, Bl. 326, 330.

position, he explained, could become very important for the "future development of the German physics" or it could recede into "utter insignificance." He argued that

> If a really productive mind [*Kopf*] does not come to Göttingen, the many younger workers who gather there like hardly any other university will not be directed in the right way, and in particular the many suggestions, which there are like nowhere else, will remain unused by the mathematical and theoretical-physical side. Naturally, there will also be a lack of feedback from experimental physics to theory and mathematics, which are particularly desirable in Göttingen. For in the development there, the mathematical ones currently outweigh the experimental ones to such an extent that they have, one could say, become almost autocratic against their will.[25]

This unequivocal assessment points to two distinctive characteristics of the development of physics in Göttingen in the 1910s. First, the guiding role of mathematics for the development of physics is noted as a source of inspiration for theory. Here Wien's term "autocratic" might be read as the turning to criteria of inner consistency and notions of simplicity that motivate expectations for physical relationships. Second, it draws attention to the fact that not only unfamiliar mathematical reasoning and content, but also a special group of "younger workers," e.g. mathematicians, infiltrated physics in Göttingen. Thus Wien's comments both realize the extraordinary energy of young talent and foreshadow, however deprecatingly, theories like the later Göttingen matrix mechanics. Further, in his letter, we find Wien drawing a most alarming scenario:

> In my opinion the appointment of a young physics mediocrity [...] would be the worst. Then physics in Göttingen would be paralyzed for more than a generation. [...] It is my conviction that this would occur if one of those gentlemen particularly recommended by some parties, viz. Franck, Pohl, Edgar Meyer, were to be appointed at Göttingen.

Considering Stark to be the only scientifically sound candidate, he writes,

> [...] I can only regret that the appointment of Stark, whom I consider by far the most appropriate candidate despite his personal shortcomings, seems impossible for personal reasons. But if the appointment of Stark is truly ruled out, it will still be better to take a physicist who is not quite so young, but can offer positions to younger researchers, than to call one who is completely inexperienced in organizational matters and who eventually will leave the imprint of mediocrity on the Göttingen institute for four decades.[26]

It was to come differently, and some of the personifications of mediocrity would eventually become Nobel winners like Stark. The Göttingen faculty was finally supported in its decision against Stark by the ministry after it found out that even strong supporters of Stark would not like to have him in their own laboratory.[27]

When the ministry communicated that it considered the negotiations with Paschen to have failed, at the same time it explicitly asked for opinions about two other candidates not mentioned by the Göttingen faculty. This is an interesting move that demonstrates how differently candidates might have been chosen. In proposing Wolfgang

---

[25]Document 11 (p. 101).

[26]Document 11 (p. 101).

[27]Cf. Rubens' attitude in Document 13 (p. 102).

Gaede from Freiburg, the ministry shows that it was by no means clear by 1915 that quantum physicists were the superior choice. Proposing the Berlin *Privatdozent* Robert Pohl, who had explicitly been excluded by Wien, it indicates that the close interaction of theory and experiment was not necessarily seen as worth preserving.[28]

Entering a Quantum Condition

On the new list of July 1915, Robert Pohl was rated second, Edgar Meyer first and Peter Paul Koch third. When the dean of the Göttingen faculty reported on the selection procedure and the candidates' talks, which could only been arranged with difficulty, all three physicists were characterized as being interested and competent in the field of quantum or atomic physics. As the candidates should represent "pure physics in the sense of W. Weber and E. Riecke" and show off to advantage "highly topical fields of molecular physics and radioactivity, especially the ones so successfully pursued by the latter," Meyer qualified through his work on "radioactive oscillations," which allowed the number of atoms per mole unit to be determined, and he won the lead through his work on the photoelectric effect, which allowed him to "penetrate even more deeply into atomistic phenomena."[29] Pohl, for his part, was said to have a "gift for precision physics," was seen to have opened with his work on the selective photo effect "a new avenue to the solution of questions about the structure of atoms that are currently in the center of interest."[30] Even Koch, who according to the judgment of his teacher Sommerfeld was neither theoretical nor mathematical physicist, "but only a brilliant physical technician,"[31] was considered due to his work on specific heats and the Zeeman effect, although the Göttingen candidate Simon (like Max Wien and Jonathan Zenneck) was disqualified for being a "technical physicist" instead of a "pure" one; as to Gaede, only doubts of his appropriateness for the position were mentioned.[32]

At this point one can safely state that a shift in focus towards quantum physics took place, and with it a reorganization process of the Göttingen physics staff. The fact that—as a consequence of all the redefinition and trading of positions—Riecke's chair of experimental physics became Born's theoretical one, while Voigt's theoretical chair became Franck's position in experimental physics, appears to be more than a curiosity. Rather, it hints at evidence of strong dynamics on the level of staff, or personnel resources, which suggests a closer relationship with the disciplinary

---

[28]Elster to Curator, 2 July 1915, GStA PK Rep. 76 V a, Sekt. 6, Tit. IV, Nr. 1, Bd. XXIV, Bl. 181.

[29]Dean to ministry, 18 December 1915, GStA PK Rep. 76 V a, Sekt. 6, Tit. IV, Nr. 1, Bd. XXIV, Bl. 348–354: The list consisted of 1. E. Meyer—Tübingen, 2. Pohl, P. P. Koch—Munich, while Simon was not included. Instead, applied and technical physics was preferred (p. 350).

[30]Ibid.: "Begabung für Präzisionsphysik [...] der einen neuen Weg eröffnen dürfte zur Lösung der gegenwärtig im Mittelpunkt des Interesses stehenden Fragen nach der Struktur der Atome."

[31]Sommerfeld to Wien, 1 June 1916, DM Wien Papers, box 010. "Auch Wagner ist theoretischer Physiker, nur nicht die Spur mathematischer Physiker. Koch ist wohl beides nicht, sondern nur glänzender physikalischer Techniker."

[32]On Koch: "Präzisionsphysik im Sinne Röntgens," spezif. Wärme, Zeeman-Effekt;—Zweifel an: Füchtbauer—Leipzig, Gaede—Freiburg, Ladenburg—Breslau, Franck—Berlin.

constellation and with the various visions and plans that key players may have tried to realize.

As the Göttingen faculty was not interested in Stark, he tried in vain to succeed to Riecke's position after 1914, causing a "bitter dispute" that lasted until 1917 (Hermann 1970).[33] In the end, it was mainly up to Hilbert and Debye to fill the vacancy, who considered Pohl and Franck and decided for Pohl, because he promised to be a better lecturer.[34] Although things are less uniform for the "younger staff" that was considered for the associate professorship in 1915/16, a clear guideline for the arguments can still be made out. Edgar Meyer, Robert Pohl, and Peter Paul Koch cannot count as quantum or atomic physicists to the same degree as Debye or Born. The reasons that were given for their proposal, however, seem suggest the opposite.

One can leave aside the question of the extent to which the above characterizations included wishful thinking, and to what degree these candidates could be called quantum or atomic physicists, the central point is how the Göttingen faculty argued. Interestingly, a closer look at the lists of candidates for the main physics chairs, especially at times of reorganization in 1914 and 1920/21, shows that researchers in the field of quantum physics (particularly on the first list) and of atomic physics (particularly on the second) were not only first, but often the only choice. No case was found where advocates of the quantum had to compete with scientists from other research fields; classical fields like those Gaede stood for, and which had been still represented by Riecke and Voigt, had ceased to play a role.

The by this time long-standing wish to replace Riecke by Wilhelm Wien can be seen at least as a sign for a turn towards accepting quantum physics, for Wien's name and work was closely linked with the quantum problem. As all candidates of the list from December 1914 were important figures in the development that finally led to quantum mechanics—Wien, as already mentioned, Paschen as his competitor and the experimental alter ego of Sommerfeld and his theory, Stark and Zeeman as those who raised central problems of the quantum description of atoms—, it is fair to assume that candidates in this field were chosen intentionally. And Debye was clearly a representative of quantum physics, whose appointment was meant to strengthen this particular field in Göttingen. He engaged in quantum research both theoretically and experimentally. When he left and Born and Franck took over his duties in 1921,

---

[33] For Voigt's views on Stark, see his review of Stark's book on "Elektrische Spektralanalyse chemischer Atome," Leipzig 1914 (Voigt 1915a), where he concludes: "Das Einzige, was bleibt, und was für Leser, die Starks eigenartige wissenschaftliche Persönlichkeit aus seinen Originalabhandlungen nicht zuvor kennen gelernt haben, ist der Einblick in dessen eigne Vorstellungswelt. Wie bei anderen experimentell ungewöhnlich begabten Forschern ist in Stark das Bedürfnis nach lebendiger innerer Anschauung der von ihm behandelten physikalischen Vorgänge äußerst ausgeprägt, ein Bedürfnis, das sich zur Not auch mit einem Modell befriedigt, das nach manchen Seiten hin physikalisch unmöglich erscheint, wenn es nur nach der einen ihm momentan wichtigsten Seite hin eine Deutung des Vorgangs anbietet. [...] Stark ein Experimentator erstern Ranges [...]" (p. 502).

[34] Debye interview 1962 p. II/5 (AHQP). It appears that Hilbert had convinced Debye of Stark's inappropriateness. "[...] Hilbert insisted that that as a *völkisch* nationalist and outspoken anti-Semite, Stark was simply unacceptable for Göttingen, and after he apparently persuaded Peter Debye of this, no one was prepared to argue otherwise" (Rowe 1992, 502).

it had to be seen as a continuation of a research program rather than the onset of a new one.

By 1915 work on quantum physics and interest in the structure of the atom had become practically indispensable for candidates of the Göttingen physics faculty. The shift of interest away from Voigt's and Riecke's research programs, and in particular from their pessimism about quantum theory, was already complete at this point. Neither Riecke's wish for a "fresh person" nor Voigt's request that the physics institute under Debye should live up to the standards set by Franz Neumann in the middle of the 19th century[35] give any indication that they had any influence on this shift of focus. Who else could have? The "personal" full professor Wiechert may have to a certain degree, the applied physicists Simon and Prandtl definitively were not. Hence, we cannot but consider the mathematicians.

[35]Voigt to Naumann, 20 May 1914, GStA PK Rep. 76 V a, Sekt. 6, Tit. IV, Nr. 1, Bd. XXIV, Bl. 71–73. "[...] der Universität ein theoretisch-physikalisches Institut zu schaffen, das im Sinne des Ideales meines Lehrers Franz Neumann die breiteste Berührung zwischen Theorie und Beobachtung vermittelt."

Chapter 3
"Hilbert and Physics"—Vision and Resources

Klein's agenda for an applied mathematics materialized at the beginning of the century in the *Göttinger Vereinigung zur Förderung der angewandten Physik und Mathematik*, a group of professors and industrialists that tried to promote applied sciences and which raised considerable financial means for this aim.[1] This gave Klein a major influence on the appointments and institutional resources of Runge, Simon, Prandtl, and to a lesser extent also Wiechert. Except for the last of these men, none of the scientists mentioned is of greater relevance for our question.

Klein himself, the influential science organizer and adviser, did not entertain any interest in the quantum, and besides this he lost most of his influence after he fell ill in 1911. Therefore simple reasoning by exclusion suggests that it may have been David Hilbert who took over Klein's role on the new stage of opening mathematics for fields like quantum physics and atomic theory. To give further indications for the validity of this thesis, one can quote from Debye's recollections, which point to Hilbert's vigorous engagement. When it came to the choice between the two young experimental physicists Franck and Pohl, Debye remembered: "I told Hilbert 'Well you'd better look at them.' So I went with Hilbert to Berlin and we went to Ruben's laboratory—didn't say why, you see—and met both of them."[2]

Max Born's article on "Hilbert and physics," written to commemorate his sixtieth birthday, is perhaps the best account of Hilbert's physics activities, as it describes and assesses his interests and contributions to physics at a time when the quantum riddle was still largely unsolved and nobody would have anticipated that eigenvalues, integral equations and infinitely dimensional spaces, later known as *Hilbert spaces*, would become the tools for its solution. Though Born did not mention that Hilbert had studied some physics at Königsberg, and even left out the sixth problem of

---

[1] For the aims and successes of the Göttingen Association for the promotion of applied physics and mathematics, see Klein (1908) and Tobies (1991).

[2] Interview Debye 1962, p. II/5 (AHQP).

© The Author(s), under exclusive license to Springer Nature Switzerland AG 2019
A. Schirrmacher, *Establishing Quantum Physics in Göttingen*,
SpringerBriefs in History of Science and Technology,
https://doi.org/10.1007/978-3-030-22727-2_3

his 1900 Paris address, he described very clearly what he had witnessed: the 1905 seminar on electron theory "that withdrew the two befriended mathematicians from their actual field"[3]; and after Minkowski's death, the application of integral equations to the kinetic theory of gases and to elementary radiation theory. He also pointed to at least two different phases in which Hilbert had dealt with established physical theory. In the first he attempted to improve its foundations, in the second he turned to "modern" physics like statistics and quanta. Born also mentioned, that while for the latter there was no materialization into published work, on other topics he did have great local influence on students and faculty. Part of this influence is found in Hilbert's radiating verve: "The almost youthful fervor with which Hilbert entered the completely new fields of radiation theory and quantum hypothesis, thoroughly new to him, had something tremendously inspirational for everybody who witnessed it."[4] How did this come about?

Hilbert's views on the development of mathematics and his axiomatic program prompted him to look at physics, most prominently demonstrated by the sixth problem of his 1900 Paris speech and his 1905 lecture course on the "logical principles of axiomatic thinking," in which theories of mechanics, thermodynamics, electrodynamics and even psycho-physics were used as examples.[5] Hilbert's collaboration with Minkowski was on the application of the axiomatic method to rich examples of theories that were essentially considered finished by their respective fields. As Leo Corry observed: "In Hilbert's view, the definition of systems of abstract axioms and the type of axiomatic analysis described above was meant to be conducted retrospectively for concrete' *well-established and elaborated* mathematical entities" (Corry 1997, 115). Physics, as the "mother of all sciences,"[6] offered the best source of examples for the axiomatic method around 1905.

Hilbert's later writings on the axiomatic method, however, show a strikingly different attitude. In his private notebooks there is repeated emphasis on understanding the axiomatic method in the correct way and in particular against the need of having a fully developed theory before axiomatics could be applied.[7] By 1917 the axiomatic method had matured into an axiomatic thinking. It is in this new sense of a "leading

---

[3]In Born's words: "Die Anregung zu diesem Unternehmen, das die beiden befreundeten Mathematiker von ihrem eigentlichen Arbeitsgebiet abzog und sie zum tieferen Eindringen in die Nachbarwissenschaft veranlaßte, mag damals von Minkowski ausgegangen sein [...]" (Born 1922a, 88).

[4]Runge (1949, 149), the original reads: "Das geradezu jugendliche Feuer, mit dem er in die für ihn ganz neuen Gebiete der Strahlungstheorie und Quantenhypothese hineinstürmte, hatte für alle, die es miterlebten, etwas ungeheuer mitreißendes".

[5]This part of Hilbert's work has been widely discussed in the literature. A comprehensive account was given in Corry (1997), where the 1905 lectures are discussed in detail.

[6]Hilbert to Sommerfeld, 29 July 1906, DMA HS 1977–28/A,141.

[7]In Hilbert's undated private notebooks, section "Allgemeine Mathematisch-Philosophische Bemerkungen," UA-Gö, Hilbert Papers, 600:3. The original reads: "Ich protestire gegen den Einwand, die Physik sei noch nicht weit genug, zur Axiomatisierung. Jede Wissenschaft ist zu jeder Zeit nicht nur reif genug sondern erfordert mit Notwendigkeit die Axiomatisierung, diese im richtigen Sinne verstanden."

part in the sciences" that mathematics has its "calling."[8] Born described Hilbert's method as a "methodological requirement," which guides theory construction rather than reconstruction (Born 1922a, 91).

With regard to this general turn of perspective in the axiomatic method, it may suffice here to point to four basic factors that contributed to this turn. First of all, Hilbert was following Poincaré's lead. The Frenchman had addressed the problem of discontinuity, which Lorentz had raised in 1908, in his last papers written shortly after participating in the first Solvay congress in 1911. He concluded that "[...] the physical phenomena cease to obey laws expressible by differential equations and this will be, without any doubt, the greatest revolution and also the deepest one that natural philosophy has gone through since Newton" (Poincare 1912, 626). The leading mathematicians Hilbert and Poincaré, though in competition from time to time (e.g., on geometry or integral equations), were always cordial to one another and both became proponents of quantum theory. Upon Poincaré's death Hilbert might have felt an obligation to continue Poincaré's ideas and probably saw an opportunity to bring Göttingen into play.

A second factor was the advent of atomic models like Thomson's, Rutherford's, and Bohr's, which extensively employed new rules and hypotheses in a way that suggested improvement by axiomatic formulation. Third, quantum theory in general was implemented into physics as a kind of additional, axiomatic hypothesis. The axiomatist, therefore, clearly saw the all-important need to avoid contradictions, which could falsify the entire framework. And finally, even general relativity and Mie's related theory gave rise to an attempt for an axiomatic treatment by Hilbert, which he communicated in papers and lectures on the "foundations of physics" (Hilbert 2009a).

Hilbert's turn in axiomatics coincided with a turn in quantum theory, which shifted from considering the problems of black-body radiation to the application of the quantum to atomic structure.[9] As Arthur Haas' physical model for Planck's resonator, which employed Thomson's atomic model, stood at the beginning of this process (besides ideas of Thomson himself and Nicholson), its discussion by Lorentz in his 1910 Göttingen lectures may be seen as a further ingredient in the Göttingen perception that the quantum was becoming relevant to atomic constitution. In this way, quantum physics already became popular in Göttingen circles even before the 1911 conferences in Karlsruhe and Brussels put the topic more visibly on the agenda. As the historical analysis of early quantum theory has used publication analysis to show "that by 1911 and 1912 the quantum had entered physics to stay and that a major reconstruction of classical theory had become inevitable," with Hilbert's example we have found a case that demonstrates how such a shift in focus towards quantum problems can be found and explained on a local level.[10]

---

[8]Ibid. "In dem Zeichen der axiomatischen Methode erscheint die Mathematik berufen zu einer führenden Rolle in der Wissenschaft."

[9]For the following cf. Kuhn (1978, 206–232).

[10]Cp. Lorentz (1910), with annotations by Born, esp. p. 1252; Kuhn (1978, 228).

It is instructive to recognize that the case of the development of quantum theory differs to a great extent from that of relativity, which was largely taking place at the same time. Besides Einstein, Hilbert also contributed to the successful formulation of general relativity, in particular in 1915 (Hilbert 1915; Sauer 1999). However, while from Hilbert's perspective the hopes to find new mathematics lay in the field of quantum problems, the mathematics used for general relativity appeared to him as of 19th-century origin. As Debye recollected, Einstein had to learn the right mathematics in order to succeed with his theory; for mathematicians, and in particular for Hilbert, "everything was finished with Minkowski," since the mathematics of general relativity "was something quite common [gang und gäbe] [...] for mathematicians it was nothing especially new."[11] Accordingly, Klein had also tried to subsume general relativity in his Erlangen program (Frei 1985, with comment on p. 142). As a consequence, the development of general relativity—as important it may have been for the conceptual progress of physics—appears surprisingly neutral with respect to the question of the impact the programs of mathematicians might have had on the local development of physics in Göttingen.

3.1 Hilbert Investing in Physics—Personal Resources

The central question of this study concerns the effects that occur when scientific interest becomes transformed into action with respect to resource allocation and research politics. The development of Hilbert's strong interest in recent physics and its resonance in the mathematics community will be complemented in the following with the analysis of the recruitment and use of assistants, the teaching and training of doctoral students, the discussion of physics topics that took place in the *Mathematische Gesellschaft*, with Hilbert's lecture topics, and with the spending of additional funds, in particular the Wolfskehl prize money. All of these fields can be understood as personal resources because they constitute assets that a person can influence or gain knowledge from. This can happen either by making one's own decisions about what to work on, by delegating or directing work in specific directions, by organizing forums for information exchange, or even by sitting on a committee that distributes funds. In this way, I will try to assess the role of mathematicians in the reorientation of physics research. I will ask more directly, to what extent David Hilbert became a kind of *Ersatz* physicist, albeit only temporarily, when he was both filling a vacuum in advocating for modern physics and pursuing genuine aims of fostering modern mathematics.

The most obvious resources for Hilbert's research were his assistants, with whom he often established an intimate working relationship, mostly at his home. Some of them digested and reported progress in mathematical and physical matters to Hilbert in regular meetings, others took notes of his lectures and revised them with their master later in the day; Hilbert preferred to talk science, rather than to read

---

[11]Debye Interview 1962, p. II/13 (AHQP).

it. Although the assistants followed individual lines of research, most of them were closely related to their master's interests. It is revealing to see the extent to which mathematics and physics assistants were distinguished, both financially and with respect to their importance.

Table 3.1 collects the available information on Hilbert's assistants as reconstructed from the files of the Ministry of Culture and from correspondence.[12] First of all, it turns out that Hilbert had various types of assistants. We find the unpaid *Privatassistent* like Max Born, who did not depend on financial support but benefited from personal contact with his master while discussing the lecture notes to be made available in the mathematical reading room.[13] Similar was the case of Paul Peter Ewald, who started his studies with chemistry at Cambridge but turned quickly to mathematics at Göttingen. He, however, was paid privately by Hilbert.[14] Next there is the group of regular assistants who, after having taken their doctoral degrees, were paid moderately from funds provided by the state. This was the case for Ernst Hellinger, Alfred Haar, Richard Courant, Wilhelm Behrens and Erich Hecke. The latter may also fit into a group of young up-and-coming professional academics who were assistants in order to receive some support, while otherwise working as unpaid lecturers (*Privatdozent*). In general one observes an interesting pattern of maturation in Hilbert's assistants over time.

The physics assistants can also be accommodated in this scheme. They appear as a second regular type of assistants, mostly postdoctoral researchers of different backgrounds. Ewald initially came from chemistry and had first turned to mathematics (under Hilbert) and then to physics (under Sommerfeld in Munich). Alfred Landé, who started as an experimental physicist like Erich Hückel, had turned to theory.[15] Furthermore, there were additional students and researchers of comparable standing close to Hilbert and his work. In this group one might also put Otto Toeplitz, who had collaborated a great deal with Hellinger on a theory of infinite-dimensional matrices, a topic Hilbert had raised and which became a central resource in the formulation of matrix mechanics,[16] and Hermann Weyl, who did some mathematical work related to black-body radiation.[17]

As regular paid positions existed only for the applied branches of mathematics, represented by Runge and Prandtl from 1905 on, and for the collection of

---

[12]"Allgemeine Angelegenheiten des mathematisch-physikalischen Seminars 1869–1923," GStA PK 76, Nr. 591, pp. 176–300, in particular Hilbert to Elster, 8 March 1907 (182–183v), and Hilbert to Wende, 17 February 1920 (260–262).

[13]Born had regular contact with Hilbert from summer 1904 on and prepared lecture notes. Due to family relations to Minkowski he was more friend than student of both. From 1905 on he was appointed *Privatassistent*. Due to the relative wealth of his family there was no need to get a paid position, in contrast to e.g. Hellinger, who depended on support. Cf. Born (1975).

[14]According to Reid (1970, 108f), Ewald served as "a special *Ausarbeiter* for a large class" and received a low salary from some non-mathematical funds Hilbert had acquired.

[15]Hückel was later also assistant with Born and Debye.

[16]Toeplitz took over a professorship in Kiel in 1913 but continued work with Hellinger, cp. Toeplitz and Hellinger (1906, 1910, 1927).

[17]Weyl, however, was strictly addressing a mathematical audience, cp. Weyl (1912a, 1912b, 1913).

Table 3.1 Hilbert's assistants and their pay 1904–1930, compiled from various sources

	Mathematics assistants	Pay/yr.	Physics assistants	Pay/yr.
(SS 04, WS 04)[a] from SS 05	Born Privatassistent	—		
WS 05 SS 06, WS 06	Hellinger	900[b]		
(WS 06)	Ewald	?		
SS 07, WS 07 SS 08, WS 08	Haar	900		
SS 09, WS 09 SS 10, WS 10	Courant (1910: Dr.)	900		
WS 10	Behrens	1200		
SS 11, WS 11	Hecke (Dr.)	1200		
SS 12, WS 12	Hecke (PD)	1200	Ewald (Dr.)	800[c]
SS 13, WS13	Hecke (PD) as 1st assist. (Baule)	1200	Landé as 2nd assist.	800
SS 14, WS 14	Hecke (PD) as 2nd assist.	1200	Landé (Dr.) as 1st assist.	1200
SS 15, WS 15 SS 16 (part)	Krafft (PD)	1200		
SS 16 (part) WS 16, SS 17	Bär (Dr.?)	1200		
WS 17 SS 18, WS 18	Bernays (Dr.)	1200		
SS 19, ZS 19 WS 19	Bernays (PD)	1200	Baule (Dr.) SS, Sponer (ZS, WS)	1200
SS 20, WS 20	Bernays (PD)	2400[d]	Kratzer (Dr.)	3000[d]
SS 21, WS 21	Bernays (PD)	10,000	E. Hückel (Dr.)	8400
SS 22 ... 1927	Bernays (a. pl. Prof.)	(16,000)	Nordheim (1923: Dr.)	(16,000)
After 1927	Bernays, Schmidt, Gentzen		Wigner	

[a] Periods of no official appointment but collaboration in annotating lectures etc. are given in brackets. SS = summer term, WS = winter term, ZS = special additional terms after World War I
[b] Hilbert supplemented M 300 to the M 600 provided by the ministry for SS and WS 06
[c] According to Reid (1970, 130), M. 50 per month only. No evidence was found for this fact or for the doubling of this sum also mentioned upon a visit by Hilbert to the ministry in its files
[d] Sums appropriated; Hilbert requested M 3000 for Kratzer and 2–3000 for Bernays; later payments adjusted for inflation differed

mathematical models and the mathematical reading room, strictly speaking, none of Hilbert's assistants belonged to the personnel of the university as it was listed in the university catalog. They were assistants to Hilbert only, and they had been granted by the ministry *ad personam*. Some of them, however, came from one of the institutional assistantships or received one later. For example, the position of "assistant of the collections of mathematical instruments and models," was held, among others, by Hellinger (winter term 1907 to winter term 1908), Hecke (summer and winter term 1910) and Behrens (summer and winter term 1911). Hertha Sponer, in turn, came to James Franck after having been Hilbert's physics assistant, as did Erich

Hückel who then worked for Born, thus again exhibiting the close relations between physics and mathematics researchers on the assistant level.[18]

The table of assistants reveals another, rather different feature of Hilbert's recruiting politics. It strikes the eye that, at least before World War I, a quite homogeneous group of assistants was selected by Hilbert as regarding age, geographical and religious provenance. Max Born (born Breslau 1882, school and studies there), with Breslau-born or educated Toeplitz (1881 Breslau, school and studies there), Hellinger (1883 Striegau, school and studies at Breslau) and Courant (Lublinitz 1888, school and studies at Breslau), recalled that "it was quite natural that we four from Breslau constituted a group and that the others recognized us as a group. Particularly Hilbert did so very consciously" (Born 1975, 138).[19] This group of central European Jews, which was as the first, or at most the second, generation with access to academic careers in certain fields, especially mathematics, also included Alfred Haar (Budapest 1885) and to lesser extent Paul Ewald (Berlin 1888, Jewish wife and ancestors).[20] The impact of Judaism on career paths and membership in Göttingen groups, as well as in scientific fields, methods and problems—as superposed to the disciplinary development of the mathematical sciences—cannot be neglected. In this respect Göttingen was a particularly welcoming place, although there were limits (Rowe 2009).[21]

As the result of a meeting in January 1912 in the Berlin ministry, Hilbert was now able to appoint two assistants instead of one from summer term 1912 on, one of them specifically for physics. This was a concession to keep Hilbert in Göttingen, who had received a number of offers from other universities, also from abroad. For recruitment Hilbert collaborated with Sommerfeld and a typical Göttingen-Munich-Göttingen exchange was established.[22]

---

[18]Cf. *Amtliches Verzeichnis des Personals und der Studierenden der Kgl. Georg-Augusts-Universität zu Göttingen.*

[19]Born, Courant and Hellinger were inspired by the same teacher of the Breslau König-Wilhelms-Gymnasium, cf. Staley (1992, 29–36).

[20]One could probably define a second group of north German Jewish mathematicians and physicists who were influenced and considerably supported by Hilbert, including among others Hermann Weyl, Alfred Landé and Paul Hertz. One might also wonder why Paul Ehrenfest, although present at Göttingen for a considerable time, did not find entry to this congenial group; he was called a "representative of the Göttingen school," see Ioffe (1967, 42).

[21]Born was not able to get a position for Otto Stern, with whom he had collaborated closely in Frankfurt before his return to Göttingen in 1921, neither in Frankfurt nor in Göttingen.

[22]Hilbert to Naumann (*Ministerialdirektor* in the Prussian Ministry of Culture), 10 March 1912, GStA PK Rep. 76, Nr. 591, Bl. 204, where it reads: "Sodann haben Euer Hochwohlgeboren mir—gelegentlich meines Besuches Anfang Januar—einen zweiten Hülfsassistenten zu 800 M. für meine mathematisch-physikalischen Studien bewilligt. Ich habe dafür einen Schüler von Sommerfeld—München gewonnen, nämlich Dr. Ewald—München, [...]" Since nothing can be found about this meeting in the papers of the ministry, one can only speculate that Hilbert used, as he frequently did before, offers of positions elsewhere to improve the Göttingen situation. (At this time Carl Neumann retired from his Leipzig chair, the position was ultimately not filled, but converted into an associate professorship for Göttingen *Privatdozent* Paul Koebe; among the other professors were Otto Hölder and Gustav Herglotz. Also, Philipp Lenard wrote to Sommerfeld on 25 September 1913 of a failed attempt to win Hilbert for a new second chair of mathematics in Heidelberg. DMA HS 1977–28/A,1982).

Ewald was Hilbert's *Privatassistent* and he shared a flat with Born, von Kármán and other students of Hilbert during his first Göttingen period. He was taking notes on one of Hilbert's lectures and perfected them under his direction and, in this way, was part of the distinctive Göttingen crowd, which was particularly open for new mathematics and its neighboring fields (Kaiser 1970; Born 1975; Hilbert 1907). After completing his doctorate with Sommerfeld in Munich on "the dispersion and birefringence of electron lattices (crystals)" (Ewald 1912), he returned to Göttingen with an option for habilitation.[23] When he joined Sommerfeld a year later, he would be replaced by Alfred Landé, another good example for collaboration by sharing students: After having studied in Marburg and Munich for two years, he continued in Göttingen from 1910 to 1912 and began his dissertation under Sommerfeld in Munich thereafter, but went as assistant to Hilbert in spring 1913. He presented his dissertation "On the method of eigen-oscillations in quantum theory" (Landé 1914), which was published in Göttingen, to the Munich faculty in the middle of his Göttingen period in 1914.

The war made Landé leave his Göttingen position, however, as he wanted to do his duty as a paramedic in the Red Cross, reportedly to some annoyance on Hilbert's part.[24] His case offers a good indication of what Hilbert wanted. In an interview Landé recalled "I went to Göttingen already a convinced quantum theorist." and described his duties: "Every morning and afternoon I had to report to Hilbert on new literature in quantum mechanics [NB: quantum physics is meant], on ideas about the behavior of solid bodies at low temperature, on spectroscopy, and the like."[25]

The fact that a mathematician spends part of his financial means on the borderland to a neighboring discipline, rather than for central mathematical questions, may not be too surprising. However, Hilbert's actions went much further. Not only did he employ researchers of a standing the physics institute should have been striving for, they—and what may be even more extraordinary, even he himself, as we will see below—were dealing with genuine physics problems, work that could have earned them advanced degrees elsewhere. Moreover, the physics assistant was not only equal to the even more highly qualified mathematics assistant in financial terms, he also was ranked higher. Hilbert wrote to the Ministry in spring 1913:

> For the forthcoming budgetary year I have employed my past 2nd assistant Dr. Landé (theoretical physics), who has become indispensable for the preparation of my theoretical physics lectures, as 1st assistant, and thus an increase in his remuneration from 800 Marks to 1200 Marks appeared necessary to me.

---

[23] Hilbert to Naumann (*Ministerialdirektor* of the Prussian Ministry of Culture), 23 April 1912, GStA PK Rep. 76, Nr. 591, Bl. 208. It reads: "Ich bemerke noch, dass Dr. P. Ewald ein aufgebildeter theoretischer Physiker ist, der beste Schüler von Sommerfeld—München und künftiger Habilitand, der mir in Korrespondenzen und Besprechungen—er ist bereits hier in Göttingen—schon jetzt die besten Dienste geleistet hat."

[24] Cf. Reid (1970, 141); the letters from Landé to Hilbert of 1915, however, suggest a less dramatic account. UA-Gö, Hilbert Papers, folder 207.

[25] Interview Landé, 1962, session I, pp. 7 and 5 (AHQP).

The mathematical assistant had only a narrow escape from financial demotion, as Hilbert cleverly proceeded

> My past 1st assistant Dr. Hecke, *Privatdozent* for mathematics, will dedicate his assistant service to me one more year and thus shall become my second assistant. Since Dr. Hecke so far received 1200 Marks and gave invaluable service, I propose to appropriate for the forthcoming budgetary year as an exception 1200 Marks instead of 800 Marks, which is what my second assistant usually gets.[26]

This exceptional instance shows a clear turn of Hilbert's investments from fostering mathematics to taking over physics.

In later years, and in particular in the period of World War I, great changes occurred with respect to the availability and tasks of the assistants. Here we find a second instance of inverted investment in Hilbert's request for remuneration of Paul Bernays and Adolf Kratzer in 1920. Hilbert wrote to the ministry that

> Dr. Kratzer, up to now full assistant at Sommerfeld in Munich, shall become my mathematical-physical help in the future, of whom I expect a lot. He told me that he could not come if he was paid less than 3000 Marks a year [...]

This was the same amount he also requested for Bernays, who was already *Privatdozent*. However, he concluded: "In any case I ask to appropriate for Dr. Kratzer at least 3000 Marks, for Dr. Bernays at least 2–3000 Marks for 1920/21."[27] In consequence, the analysis of expenditure on assistants produced an indicator that points to a period between 1912 and ca. 1920 in which physics had more value to Hilbert than mathematics.

And finally, Hilbert's role as a substitute physicist is also evident in the success of graduates in the field of mathematical physics (or physical mathematics), shown by the dissertations he supervised. As a number of scientists who played an important part in physics later on did their doctoral work on pure mathematics before 1911,[28] between 1912 and 1921 five of ten dissertations were related to physics: atomic structure,[29] dilute gases,[30] electrolysis,[31] and even quantum theory,[32] another one was also on philosophy,[33] but only a minority of four remained for genuinely mathematical topics.

---

[26]Document 7 (see Chap. 6).

[27]Hilbert to Ministerialrat Wende, 17 February 1920, GStA PK Rep. 76, Nr. 591, Bl. 260–261.

[28]E.g. dissertations supervised by Hilbert, written by Weyl on Fourier's integral theorem in 1908 and by Courant on Dirichlet's principle in 1910. Cp. "Verzeichnis der bei Hilbert angefertigten Dissertationen" in Hilbert (1935).

[29]Dissertation (Föppl 1912), viva voce 1 March 1912.

[30]Dissertations (Bolza 1913) dated 3 June 1913, and (Baule 1914) dated 18 February 1914).

[31]Dissertation (Schellenberg 1915) dated 24 June 1914. The theses by Bolza, Baule and Schellenberg seem to have emerged from a seminar on *kinetic theories in physics* (summer term 1913), which Hilbert gave together with Hecke, in which Born, Hertz, von Kármán, and Madelung also took part. Cf. Lorey (1916, 129), and Table 3.3 of lectures below (see in this chapter).

[32]Dissertation (Kneser 1921).

[33]Dissertation (Behmann 1918).

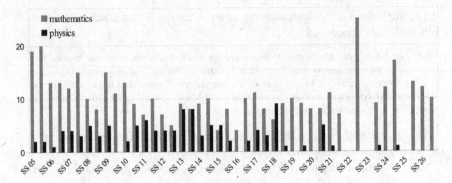

Fig. 3.1 Number of talks in the *Mathematische Gesellschaft* on topics of mathematics versus physics. Source: Lists of topics as reported in (Jb. DMV)

3.2   Physics in the *Mathematische Gesellschaft*

A further indicator of non-monetary nature of the mathematicians' investments in physics is found in the topics of the talks at the *Mathematische Gesellschaft* of Göttingen. This society was founded by Felix Klein and Heinrich Weber in 1892 and met every Tuesday evening. Similar societies existed in many university towns, though most of them were not official university institutions, but rather private-run societies or clubs.[34] Although one should keep in mind the interdependence of the various resources analyzed in this section—for example, one might expect that the physics assistants like other advanced students and researchers of Hilbert's group would very likely become speakers in the *Mathematische Gesellschaft*—the differentiation is far from redundant.[35] Looking at the talks in the *Mathematische Gesellschaft* serves to exhibit the broader impact of an elevated interest in physics topics on the part of mathematicians in general, and by Hilbert in particular.

Interpreting the Fig. 3.1 as an indicator for the spending of a special kind of resources, viz. attention, it shows a period of particular high interest in physical problems from 1910 to 1915 and in summer 1918. In the summer term 1911, and from winter 1912 to winter 1913, the number of physics talks equaled or almost equaled that of mathematical ones, while in the summer terms 1915 and 1918 physics, in fact, outnumbered mathematics.[36]

---

[34]Cf. Lorey (1916), esp. pp. 138–141 and 219–221.

[35]Actually, Ewald spoke only once at the *Mathematische Gesellschaft* on his dissertation, on 4 June 1912, and Landé not at all; Born, on the contrary, spoke a dozen times between 1910 and 1915 but only once on a topic not related to physics; in this period Hecke gave three talks on mathematics and one on physics.

[36]The summer 1911 physics talks were by von Kármán (hydrodynamics), Weyl (radiation, discussion of Lorentz paper), Born (*X*-rays), Wiechert (relativity), and Philipp Frank (statistical mechanics). The physics talks at the "plateau" of 1913 are mainly about radiation, quantum theory, atoms and atomic lattices, and relativity. In summer term 1915, talks by Einstein and Sommerfeld are to blame for the predominance of physics, as are those of Max Planck in 1918. Cf. Jb. DMV (1918).

A good example of the role of the society in exchanging news is the Göttingen discussion about the outcomes of the first Solvay Conference, which took place in Brussels in fall 1911 (without participation from Göttingen). In two sessions on February 25 and March 4, 1913, Born and von Kármán obviously gave an in-depth report on the recently published French version of the proceedings. As Born later published a brief report in *Physikalische Zeitschrift*, one can catch a glimpse of the Göttingen reception. Born staged the "Brussels quantum congress," as he once called the meeting in *Naturwissenschaften* (Born 1913a), here as a fight between two groups or "parties," the "quantum friends" and their foes. To a much greater extent than usual in the scientific literature, the individual personalities come to bear on the discourse, "for as long as the foundations of quantum theory are so barely clarified as today, the favorable or unfavorable attitude to the new theory depends significantly on personal preferences and views." The party of the quantum friends, however, was dominant, while Lorentz stood above both directions (Born 1914b, 146). In this way the Göttingen mathematics community was well-informed about the developments of the new quantum theory.

When in 1921, however, discussions on physics died down in the *Mathematische Gesellschaft*, only Born would return for single talks in winter 1923 and winter 1924, respectively, as after the advent of quantum mechanics, physics was no longer a topic. Ironically, Born's last talk on 9 December 1924 was on "atom mechanics and other things," but not on any actual advances, which occurred half a year later.[37] This incident points out a crucial fact about Göttingen's history of quantum physics: ironically, the mutual interaction between mathematics and physics was reduced precisely at the height of the search for a quantum mechanics.

Regarding the relationship between physics and mathematics as academic fields in the university, the indicators analyzed demonstrate a specific kind of disciplinary imperialism by Göttingen mathematics. The question is whether what happened was in fact a shift in disciplinary boundaries, or whether it would be more accurate to speak of a mutual penetration or overlapping of interests.[38] A conceivable test could be to juxtapose the debates of the mathematical society with those of the physical society, which also existed and met regularly. Unfortunately, no records comparable to those for the mathematicians have been found, only occasionally are talks reported or can be reconstructed from notes and recollections. The sparse information available, however, does not suggest a symmetrical relation. Neither did the physicists begin to talk mathematics, nor did the mathematicians start joining them. Hilbert, for example, presented a paper on the problem of specific heats in the physical society in January 1913, which I will discuss below (p. 42); here Hilbert carefully tried to adapt to the physicists style, for example, by comparing predictions with measurements and by giving explicit quantities.[39] Only few additional cases are known, which concerned

---

[37]Mentioned in Jb. DMV (1926, p. 103), "Atommechanik und anderes."

[38]The thesis that the border between mathematics and physics became permeable is attributed to Sigurdsson (1991).

[39]Manuscript "Bemerkungen zum Nernstschen Wärmesatz" of talk at *Physikalische Gesellschaft* on 15 January 1913, Hilbert Papers, folder 590, Bl. 1–16.

atomic models. Debye discussed Rutherford's nuclear atomic model in spring 1914, while Born presented a paper "On the stability of Bohr's atomic model" in December of the same year.[40]

3.3 Hilbert's Physics Teaching

In his revealing article "Hilbert and physics" Max Born explained that one should look at Hilbert's teaching record in order to appraise his efforts in dealing with physics. For many years he had sacrificed much time and work to deal with quanta and statistics; however, "he never published his results, but introduced them to his students in his lectures" (Born 1922a, 91).[41] To what extent Hilbert actually acted like a physicist can thus be assessed by looking more closely at his teaching at the university, but also by considering his communication with physicists, his research notes and even some publications which did in fact appear. The methodological reorientation of the axiomatization program towards topical physics problems turns out to be apparent in his lectures as well, most of which were carefully recorded by students and used by generations of students in the library of the mathematics institute and which are preserved.

The question of the extent to which Hilbert's interest in physics was concerned with quantum problems in particular has to be addressed with respect to the fields and problems, which at this time were linked to Planck's constant. Quantum physics thus was mainly radiation theory and the discussion of physical phenomena related to some fundamental discontinuity—as the title of Thomas Kuhn's book captured precisely: "Black-body theory and the quantum discontinuity." This, in fact, characterizes well what actually interested Hilbert in his teaching, for example, in his summer 1912 lectures on radiation theory.[42] However, the series of lectures on physics was longer, and those on radiation theory constituted just one step in an ongoing program. As Hilbert wrote in his autobiographical sketch:

> Decisive for my work was the closest possible connection between research and teaching. [...] It was my principle not to present in the lectures and even the more so in the seminars generally accepted [eingefahren] and as evenly as possible polished knowledge [Wissensstoff], which would have made it easy for the students to keep clean notebooks. Rather, I have always tried to illuminate the problems and difficulties and to build a bridge to the current questions. Not only on rare occasions did it happen that during a semester the program of the subject matter [stoffliche Programm] of an advanced lecture course was essentially altered, because

---

[40] Born's full paper can be found in the Hilbert Papers, 690, Nr. 1, Bl. 1–10, while there is only one page on Debye's presentation, folder 693, Nr. 2, Bl. 9.

[41] The original reads: "Auch Hilbert hat [...] seit seinem ersten Eindringen in die Gedanken der modernen Physik den Schwerpunkt der Fragestellung in den Problemen der Statistik und der Quanten gesehen. Mehrere Jahre hat er viel Zeit und Arbeit darauf verwandt, diese Gebiete mit der Schärfe seiner Logik zu durchdringen und erhellen; aber er hat seine Resultate niemals publiziert, sondern nur seinen Schülern in Vorlesungen bekannt gemacht; [...]".

[42] Hilbert (1912c), edited by Arne Schirrmacher in Sauer and Majer (2009, 435–501).

I wished to deal with topics which occupied me as a researcher and which by no means had yet gained a definite form. Advanced lecture courses of this kind resulted in a close interaction with the listeners who on their part reacted with criticism and their own ideas.[43]

Hilbert's 1911 lecture on mechanics, a standard subject in the mathematics curriculum as it may seem, is a good example of how early he extended the scope of his lectures onto physical ground. After sections on vector analysis, small deformations of continuous media, elasticity, and hydrodynamics, genuine physics topics take up more than half of the lecture notes. These are thermodynamics, the theory of relativity as well as electrodynamics and the problems arising from their combination, viz. Born's theory of the rigid body, which already refers to atomism, and relativistic thermodynamics, which also deals with black-body radiation (Hilbert 1911). Interestingly, for the same term Runge listed a course on mechanics as well, so that it may be concluded that the privilege to ignore the listed topic was mainly Hilbert's; in addition, it was up to Voigt to deal with the mathematics necessary for physics students in a course on "partial differential equations in physics."[44]

While in summer 1911 Hilbert was mostly reporting about established knowledge and some recent work of others, one of Hilbert's "most remarkable" courses of lectures followed (Born 1975, 128). His winter 1911/12 lectures on the "kinetic theory of gases" are both the attempt to reformulate a well-established physical field with a unifying mathematical framework (integral equations) and the first step towards atomic and quantum physical questions (Hilbert 1912a). He eagerly tried to show how the mathematically more rigorous framing of physical theory gives rise to criticism and hence improvement of the theory itself. In his 1912 book on integral equations and their applications to physics he employs a traditionally sober tone; in his lectures, however, a more programmatic attitude is unmistakable (Hilbert 1912b). On the very first page of the introduction he makes his standpoint clear: the phenomenological point of view must be discarded because it has no unifying power, it merely fragments the whole of physics into many single chapters, each with its own principles. Delivering these lectures to an audience of Göttingen mathematics and physics students and later spread in form of lecture notes, they criticized Voigt's well-known views and made clear that Hilbert did not see unification in Voigt's mathematical systematics.

Hilbert argued that on the basis of atomism and axiomatic formulation for the entire physics a better approach is guaranteed. This one he wanted to follow in this lectures. The best approach, moreover, were a theory of the molecular structure of matter, which was just becoming understood, and which he announced to be the topic of a forthcoming lecture, which he actually delivered in winter 1912/13 as "molecular theory of matter." To highlight the programmatic character of the

---

[43] Document 19 (see Chap. 6).

[44] For Born's characterization of Hilbert's use of lectures for new ideas cf. (Born 1975, 128). See also Verzeichnis der Vorlesungen auf der Georg-Augusts-Universität zu Göttingen, summer term 1911 ff. Voigt regularly taught on differential equations and on vector analysis. Compared over a number of years, there is considerable overlap between Voigt's and Hilbert's listed courses on the same topics: mechanics, electron theory fundamentals of physics.

Table 3.2 Hilbert's three points of view from 1911 (Hilbert 1912a)

A	B	C
"Phenomenological"	"On the basis of atomic theory"	"A theory of the molecular structure of matter"
"All of physics is fragmented into single chapters: thermodynamics, electrodynamics, optics etc." "Each field builds on specific basic assumptions"	"Single point of view for all phenomena"	"Goes much beyond B"
"*Notbehelf*," "a first step in understanding", "urgently to be left behind in order to penetrate into the actual sacred objects [eigentlichen Heiligtümer] of theoretical physics"	"Attempt to find a system of axioms valid for all physics"	? [deeper foundation, axioms = reality]
"Partial differential equations"	Mathematical method "entirely different," e.g. probability calculus, "not yet fully developed"	? [new mathematics]
? [Voigt]	"This lecture" (summer 1911)	"One of the next semesters" (winter 1912/13)

three approaches mentioned in his winter 1911/12 lectures, they are suggestively juxtaposed in Table 3.2.[45]

Obviously, the notion of axiomatic treatment is reserved for a unifying structure that captures "all physics" and it is put in parallel with atomistic theory. In this way atomistic reductionism of the physical reality, in Hilbert's thinking, was probably to be related to logical reductionism on the level of theory.

The larger aim was a theory of molecular structure of matter, a theory that not only referred to atomistic models but actually allowed all physical properties to be deduced from something even deeper than the system of axioms for all physics sought in the approach (B). But what should this be? As this scheme illustrates, Hilbert left three slots open: There is no example given for the phenomenological approach A,

---

[45]In the original the main characteristic passages read (Hilbert 1912a):

(A) Man kann die Mechanik der Kontimua rein phänmenologisch behandeln. Die ganze Physik wird dabei in viele einzelne Kapitel zerlegt [...]

(B) Wesentlich tiefer eindringend kann man die theoretische Physik auf Grund der Atomtheorie behandeln. Hier ist das Bestreben, ein Axiomensystem zu schaffen, welches für die ganze Physik gilt. [...]

In letzter Linie kann man das Hauptziel der Physik betrachten: die Theorie vom molekularen Aufbau der Materie. Dies geht noch erheblich über (B) hinaus. In einem der nächsten Semester werde ich Gelegenheit nehmen, Ihnen das, was man heute von dieser Frage weiß, ausführlich darzulegen.

which can easily be filled in with Voigt. The other two empty slots for the best of all approaches concern Hilbert's motive for dealing with physics. The scheme may suggest the answers: After a "first step in understanding" by phenomenological means (A) and a successful axiomatization "on the basis of atomic theory" (B), the objective was clearly to give a relationship between the basic notions employed in the axiomatization and the actual physical objects (C). As partial differential equations are the toolkit for phenomenologists, mathematics like probability calculus is the means of a convinced atomist. Thus in the realm C, aimed at people like Hilbert, who believed in a pre-established harmony in nature, there was the hope that advanced or novel mathematics would in some way reveal the identity of the axiomatic model and physical reality.

This specific conceptualization of the relationship between mathematics and physics, which defined a kind of physical mathematics, was pursued by Hilbert for the next couple of years, and thus established before he had acquired a physics assistant or a physics tutor. It seems that Hilbert initially planned to proceed with establishing a comprehensive theory of the axioms of physics. Hence, after having given a model by the integral equation treatment of the kinetic theory of gases,[46] he decided to deal first with the theory of radiation and quanta in a similar way, before turning to the general theory of physics axiomatization (Hilbert 2009a, 443).

It was these lectures with the title "radiation theory" in summer 1912 where the first physics assistant Paul Ewald was put to good work. They had a similar scope as the "mechanics" a year before. The emphasis was on relativity, electrodynamics and in particular on black-body radiation. In the introduction Hilbert openly showed his new interest in "real physics [eigentliche Physik], which comes from the point of view of atomism," pushing the axiomatics of mathematics and of less topical physics like particle mechanics aside, since, encouraged by the news about Laue's experiment: "one can say that no time is more favorable and more challenged to examine the foundations of this discipline than today." This was due in particular to "the atomic theory, the principle of discontinuity, that emerges more and more clearly and is longer a no hypothesis, but like the theory of Copernicus, experimentally proven fact" (Ibid. 442). Again, here we find all topics at hand that were related to quantum physics at the time!

The announced lectures on the "molecular theory of matter," however, fell short of the envisaged type C approach. As the reduction to molecular processes was still emphasized, a mixture of derivations by mathematical proofs and (phenomenological) definitions from experience was used to account for thermodynamic properties. The kinetic part dealt with ideal gases, the quantum hypothesis and the necessity to use mean values as well as the brand-new Born-von Kármán theory of specific heat. It seemed that the interest in learning as much as possible about "real physics" dominated his programmatic aims.[47]

---

[46] Planned title of lecture mentioned in Hilbert to Sommerfeld, 5 April 1912, "Prinzipien und Axiome der Physik," Document 1 (see Chap. 6), announcement in *Mathematisch-naturwissenschaftliche Blätter* as "vierstündige Gratisvorlesung" on "Grundlagen der Physik."

[47] The version of the lecture notes in the Born papers has more elaborations on quantum problems.

In January 1913 Hilbert had virtually become a theoretical physicist. He even gave a talk at the Göttingen *Physikalische Gesellschaft* (although, as we have seen, the *Mathematische Gesellschaft* had almost as many talks on physics as on mathematics that year). His manuscript on the Nernst heat theorem can hardly be distinguished from a physicist's paper, in particular as calculated values for transition temperatures were compared with measured ones.[48] He only omitted the next obvious step of publication, so that later only few recalled evidence of Hilbert's temporary conversion.

The year 1913 became the heyday of Hilbert's interest in physics, as is well documented in his lectures and papers. One should add the much publicized 1913 Congress in Göttingen and, as another more subtle example, his letters to physicists who were working on quantum problems requesting new publications in this field. For example, he wrote to Einstein, asking for his papers on the theory of gases and radiation, and to Ehrenfest who sent his latest publications, albeit, fearing that they would not satisfy Hilbert's interests, "But I did not publish anything else on quantum questions this year." he wrote.[49]

Disciplinary boundaries had ceased to play a role, at least for Hilbert and Born. At the same time the latter had pushed forward a plan with Ferdinand Springer, who may have attended the *Gaswoche*, to establish a new journal of "physical mathematics," *Zeitschrift für physikalische Mathematik*, as correspondence from May 1913 shows. It was actually a project that had already been proposed some years earlier, but had been rejected, mainly due to objections by Sommerfeld; now it did not work out either (Holl 1996, 52–54).[50] However, it still exhibits the wide scope of Hilbert's effort to link mathematics and physics.

Hilbert's teaching continued both to connect with modern theoretical physics in general and with his particular quantum interests now related to statistical mechanics. In summer of 1913 Hilbert's lectures show his interest in "real physics" topics ranging from relativity theory, application of gas theory to electrons and electromagnetic radiation, to such new subjects like electronic motion in metals and modifications of

---

[48] Manuscript "Bemerkungen zum Nernstschen Wärmesatz" of Hilbert's talk at the *Physikalische Gesellschaft* on 15 January 1913. Hilbert Papers, folder 590, Bl. 1–16. In Hilbert's words: "Eine physikalische Frage, die die Mathematiker besonders interessieren muss, ist die, wie sich die Körper am absoluten Nullpunkt verhalten, wie die Moleküle dort, vom Zustand der absoluten Ruhe ausgehend, die Wärmebewegung beginnen. Das Nernstsche Wärmetheorem formuliert hierüber bestimmte Aussagen. Ich werde nämlich erstens einleitungsweise über den wesentlichen Inhalt des Nernsttheorems sprechen, sodann zweitens einige kritische Betrachtungen dazu vorbringen, und drittens werde ich einige Bemerkungen problematischer Art mitteilen, zu denen das Nernsttheorem den Theoretiker anregt." p. 1. "[...] Berechnung aus dieser Formel führt auf die Umwandlungstemperatur 369° bei Atmosphärendruck. Die Beobachtung ergibt 368°." p. 11.

[49] Hilbert to Einstein, 30 March 1912, (Einstein CP, Vol. 5, p. 439). Reply Ehrenfest to Hilbert, 18 June 1913, Hilbert Papers, folder 91. "Ich übersende Ihnen gleichzeitig meine letzte Publikation, obwohl ich fürchte, daß sie nicht Ihrer Anfrage entspricht. Aber irgend etwas anderes über Quantenfragen habe ich in diesem Jahr nicht publiziert."

[50] It seems unlikely that Springer discussed this idea only with Born and not with Hilbert. Holl suggests that three years earlier a similar idea had been put forward without the involvement of Born, who wrote to Springer on 29 June 1913 regarding the plan for such a journal that it "was as hopeless as three years ago; then Sommerfeld turned it down firmly and I am sure that he still holds this point of view." Cited after (Holl 1996, 54).

Maxwell's equations. Unfortunately, in most cases the teaching and the presentations in the seminars cannot be reconstructed. For the seminar in the summer term 1913, however, a report by the *Mathematischer Verein*, a mathematical student corporation, is known. Listed as *Kinetic theories in physics*, but reported as *Seminar on kinetic theory of gases*, Hilbert together with Hecke (not Landé!) had the students prepare talks on dimensions and magnitudes in physics, ergodic theory, Brownian motion and its theory, Hilbertian theory of gases, dilute gases, the theory of chemical equilibrium, Sakur's papers, and so on. As participating discussants primarily Born and von Kármán are mentioned.[51]

So Hilbert finally found himself under the influence of Ewald and Landé— Göttingen students and Sommerfeld disciples at the same time—and he taught genuine physics courses on the basis of molecular gas theory or crystal structure, as in the course "electromagnetic oscillations" in the winter 1913/14 covering such topics as dispersion theory, the Zeeman effect and Planck's formula. The latter was a newly debated issue, since Einstein, together with Ludwig Hopf and with Otto Stern in 1910 and 1913, respectively, had been deriving the Planck formula without any quantum discontinuity using a simple atomic model. And Planck himself had also tried to revise his theory to eliminate non-classical behavior, a move Born commended, but not without commenting on Hilbert's successes in clearing away difficulties Planck had "not fully overcome" (Hilbert 1914a, 117).[52]

As Born had mentioned in the *Vorbericht* (preview) to the 1913 Göttingen Congress, ideas had been advanced to explain quantum theory by statistical mechanics. This was probably one of the reasons for Hilbert's seminar on statistical mechanics, which took place in summer 1914. It appeared to be a quite typical introduction to this subject. Its scope coincided roughly with with the respective *Enzyclopädie* article written by Paul and Tatjana Ehrenfest, which was focusing on the principles of Maxwell(-Boltzmann) and Gibbs. The only unexpected part may have been the presentation of applications of Gibbs' principle, including a quite concrete one on the dissociation of iodine vapor.[53]

In summary, one can say that by 1914, before the outbreak of the war, Hilbert did no longer hide anymore his teaching in the neighboring fields. In a letter to the Prussian Ministry of Culture he even declared a change in priority, since he took his "2nd assistant Dr. Landé (theoretical physics) as first assistant, who has become indispensable to the preparation of my theoretical physics lectures and since it appeared necessary to me to raise his remuneration [...]".[54]

---

[51] *Bericht des Mathematischen Vereins an der Universität Göttingen*, Sommer-Semester 1913.

[52] Cf. editorial comments in Einstein CP (Vol. 4, pp. 270–273). Kuhn (1978), Born (1913b, 501).

[53] Cf. table of content and section III of notes prepared by Luise Lange, (Hilbert 1914b; Ehrenfest and Ehrenfest 1911).

[54] Hilbert to Elster, 6 April 1914, GStA PK 76, Nr. 591, Bl. 210.

3.4   Spreading Hilbert's Gospel: Seminars with Debye, Lectures by Born and Others

In the summer term 1914 two guest professors contributed to the teaching of astronomy and physics. Besides Alfred Haar, who had earned his doctorate from Hilbert back in 1909 and was now teaching theoretical astronomy, it was Peter Debye who directly entered the field of new physics prepared by Hilbert. His lectures on atomic models geared up the Göttingen scientists for Niels Bohr's first brief visit in July 1914.[55]

As lectures, which primarily addressed students, were one arena of rather restricted interaction, seminars provided more room for exchange of ideas. For example, Hilbert had listed a seminar in winter 1912 which comprised talks on the axioms of physics, while in summer 1913, as already mentioned, talks on kinetic theories in physics were presented.[56] Hilbert's lectures on statistical mechanics of the summer term 1914 had been accompanied by a seminar as well, and at least part of this seminar dealt with molecular phenomena. Among these talks was also a contribution by Debye on electrical molecular momenta.[57] It was probably here where the famous series of seminars "on the structure of matter" were initiated, as they were called from the next term on (Table 3.3). One does not have to believe in the anecdote according to which Hilbert opened each session of the seminars on the structure of matter with the words "Tell me, just what is an atom?"[58]; however it is known that Born gave a lecture on the question of the stability of the Bohr atom in December 1914.[59]

Already in the first year of the war Hilbert's physics assistants were gone and at the same time Born was offered a position at the University of Berlin. Yet Hilbert continued his mission as *Ersatz* physicist with "lectures on the structure of matter" that followed Born's theory, which he had published in the same year as the latter's "Dynamics of crystal lattices" (Born 1915). Previously, Born had taken over Hilbert's role to bring new life to physics in summer 1914 with a course on "electron theory and the relativity principle." It was a crash course on atomistic theories of matter and electricity that was written down on over 400 pages by a student. In a similar manner as Hilbert, Born pointed out that the topics "shall not be restricted to electron

---

[55]Cf. Bohr CW (2019, Vol. 2, 331) and Bohr to Oseen 28 September 1914 cited there: "I gave a couple of short talks in the seminars in Göttingen and Munich and had many lively discussions."

[56]Verzeichnis der Vorlesungen auf der Georg-Augustus Universität zu Göttingen.

[57]Hilbert took notes on Debye's talk "Elektrische Molekülmomente" of 4 May 1914, Hilbert Papers, folder 693/1; as for another paper of Weichert(?) "Über den Einfluß der Assoziation der Moleküle auf die Temperaturɲabhängigkeit der Dielektrizitätskonstanten," 11 May 1914, Hilbert Papers, folder 719.

[58]Reid (1970, 140).

[59]Max Born's talks "Über die Stabilität des Bohrschen Atommodells," 14 December 1914, notes in Hilbert Papers, folder 690.

Table 3.3 Hilbert's physics lectures 1911–1920/21 compared with Voigt's and Debye's advanced courses, joint seminars and some related courses on quantum theory by other staff

	Hilbert	Debye	Voigt *and others*
SS 11	Mechanik der Kontinua		Kristallphysik, Magnetooptik *Born* Wärmestrahlung
WS 11/12	Kinetische Gastheorie		Geometrische Optik
SS 12	Strahlungstheorie		Thermodynamik *Wiechert* Quanten
WS 12/13	Molekularth. der Materie Axiome der Physik*		Elektrodynamik
SS 13	Elektronentheorie Kin. Theorien der Physik*		Theoretische Optik
WS 13/14	Elektromagnetische Schwingungen		Theorie des Potentials *Hertz* Strahlungs- und Quantentheorie
SS 14	Statistische Mechanik	Guest professor lectures	Kristalloptik *Hertz* Strahlungs- und
	Seminar über die Fragen der statist. Mechanik[a][#]		Quantentheorie
WS 14/15	Prinzipienfragen d. Math.	Quantentheorie[b]	Part. Differentialgleichungen
	Vorträge über die Struktur der Materie[#]		der Physik
SS 15	Vorlesungen über Struktur der Materie[c]	Kin. Theorie dielek- trischer Erscheinungen	Kapillarität
	Seminar über die Struktur der Materie[#]		
WS 15/16	(Differentialgleichungen)	Quantentheorie II Thermodynamik	Optik
	Seminarübungen über die Struktur der Materie		
SS 16	Grundlagen der Physik	Röntgenstrahlen	Elektrodynamik *Wiechert* Quantentheorie
	Vorträge über die Struktur der Materie		
WS 16/17	Grundlagen der Physik II	Ausgewählte Kapitel der mathematischen Physik	Elektronentheorie und Relativitätshypothese
	Vorträge über die Struktur der Materie		Spektroskopie
SS 17	(Mengenlehre)	Grenzgebiete der Physik und Chemie[d]	Theorie und Anwendungen des Potentials
	Vorträge über die Struktur der Materie		
WS 17/18	Elektronentheorie	Neuere Erkenntnisse der Quantentheorie	Spektroskopie der Röntgenstrahlen
	Vorträge über die Struktur der Materie		
SS 18	Differentialgleichungen	(Thermodynamik)	(Mechanik)
	Seminar über die Struktur der Materie		
WS 18/19	Über Raum und Zeit	Struktur von Spektrallinien	Optik Mechanik der Kontinua
SS 19	Denkmethoden in den exakten Wissenschaften	Ausgesuchte Kapitel der kinetischen Gastheorie	Grundbegriffe der theor. Physik
	Math. phys. Seminar über die Struktur der Materie		

(continued)

Table 3.3 (continued)

	Hilbert	Debye	Voigt *and others*
ZS[e] 19	Natur und mathematisches Erkennen	(Experimentalphysik)	Theoretische Physik Thermodynamik
WS 20	(Logik-Kalkül) Mechanik	Kristallstruktur	Part. Differentialgleichungen der theoretischen Physik[f]
	Vorträge über die Struktur der Materie		
	Kolloquium über neuere phys. Literatur[f]		
SS 20	Mechanik und neue Gravitationstheorie	Mechanik	—
			Hertz Elektrizitätstheorie
	Vorträge über die Struktur der Materie		
WS 20/21	(Anschauliche Geometrie)	—	*Hertz* Relativitätstheorie
	Seminar: Struktur der Materie		

*Seminars on physics topics conducted by Hilbert
#Not listed under Debye's name, but probably done in collaboration
() Lectures of other topics, only mentioned if no other relevant lecture occurred
[a]Part of the seminar at least dealt with molecules, cf. above
[b]Lecture notes Einführung in die Theorie des Planckschen Elementarquantums, AIP Misc. Physicists Collection: Debye, MP 56
[c]At least part of the seminar dealt with atomic models: Born: Über die Stabilität des Bohrschen Atommodells (14 December 1914) Hilbert Papers, folder 690; Debye: Das Rutherfordsche Atommodell (undated) Hilbert Papers, folder 693/2
[d]Listed title only, no lecture notes available
[e]ZS is Zwischensemester, instead of summer and winter terms for 1919/20 there were three shorter terms SS, ZS and WS
[f]Listed title only, Voigt had died 13 December 1919
Sources: Verzeichnis der Vorlesungen auf der Georg-August Universität zu Göttingen, Lecture notes prepared for the Mathematische Lesezimmer, partly in the papers of Hilbert and of Born. For a complete list of Hilbert's lectures and available corresponding documents, see (Hilbert 2004, 609–623)

theory but moreover furnish the foundation for an atomistic theory of matter in the first place" (Born 1914a).[60]

Finally, Hilbert's 1915 lectures and his seminar on the structure of matter demonstrated that Voigt's field had been taken over by others in Göttingen. This might have been tolerable since the author of the enormous volume on *Kristallphysik* was to retire before summer 1915: however, the war prevented him from going to Harvard as an exchange professor during his last term as planned; instead he continued teaching until his death in 1919.[61] The 1915 lecture notes of Hilbert's course were taken under unfavorable war conditions and are much sketchier than usual. The course appears here neither programmatic nor did it challenge Voigt's *opus magnum*. Voigt in turn made it clear that his results of many years could not be so easily reduced to a mathematically appealing lattice theory, which he found in some respects "rather unconvincing." Born's models of displaced electron clouds, he still pointed out in

---

[60]In Born's lectures in the 1920s this decisive atomistic approach is not longer pushed.
[61]Cf. vom Brocke (1981, 142, 145 f.).

Table 3.4 Topics covered in Hilbert's physics lectures 1911–1920

Mechanik der Kontinua	*Thermodynamics, relativity, electrodynamics, Born's theory of the rigid body, relativistic thermodynamics, black-body radiation*
Kinetische Gastheorie*,#	*Hydrodynamics, theory of gases of one or more atoms (all in terms of integral equations)*
Strahlungstheorie*	*Relativity, radiation theory by Kirchhoff, Planck's formula*
Molekulartheorie der Materie	*Theory of gases, crystals and radiation, Zustandsgleichung, specific heat (Debye, Born and Karman), quantum hypothesis*
Elektronentheorie	*Relativity, theory of gases applied on electrons (radiation, conduction), electron "clouds," electrodynamics and modified Maxwell's equations*
Elektromagnetische Schwingungen	*Dispersion theory, Faraday and Zeeman effect, magnetism, Planck radiation law without quantum theory*
Statistische Mechanik	*Principles of Maxwell and Gibbs*
Vorlesungen uber Struktur der Materie	*Kinetic theory of crystals (after Born)*
Grundlagen der Physik*	*Special relativity, Mie's theory*
Grundlagen der Physik II*	*General relativity, causality*
Elektronentheorie	*Special relativity*
Über Raum und Zeit	*Semi-popular lectures on relativity*
Denkmethoden in den exakten Wissenschaften	*Philosophy of mathematics and physics*
Natur und mathematisches Erkennen	*Philosophy of mathematics and physics***
Mechanik und neue Gravitationstheorie	*Mechanics, relativity*

*Published in Hilbert (2009a)

**A commented and edited version has been published as Hilbert (1992)

#This is the title of the version of Hecke's notes kept in the Göttingen Mathematisches Seminar; Born's copy of these notes has the title "Mechanik der Kontinua aufgrund der Atomtheorie," Born Papers, folder 1816

1918, could hardly account for the "immensely different behavior" of various crystalline substances (Voigt 1918, 3f).[62]

---

[62]The central passage reads in the original: "Was Kristalle mit mehreren Atomarten angeht, so erklärt Herr Born in seiner bedeutungsvollen Monographie etwaige Abweichungen von der Cauchyschen Relation durch die gegenseitige Verschiebung der verschiedenen Atomarten; er hat aber, wie mir bekannt, wenigstens bei regulären Kristallen mit nur zwei Atomarten diese Darstellung als nicht haltbar gegenwärtig aufgegeben und zieht dafür Elektronenwolken' heran, die die Atome umgeben und, als gegen sie verschiebbar, gewissermaßen noch zwei weitere Atomgattungen vertreten. [...] stimmt bei Steinsalz sehr gut [...] die Zahlenwerte von Pyrit geben einen krassen Widerspruch. Dies ungemein verschiedene Verhalten auf die verschiedene Verschiebbarkeit der Elektronenwolken zurückzuführen erscheint wenig befriedigend." Cf. also the contrary account of Voigt's assistant from 1909 to 1914 in Försterling (1951), in which Voigt's embrace of Born's theory is repeatedly emphasized.

From 1915 on, Hilbert reduced his work on questions related to quantum physics and the structure of matter, since he had entered into another physics project: general relativity. Starting from the recent ideas of Gustav Mie, he lectured in summer 1916 on "foundations of physics," which essentially meant "the modern relativistic ideas" and continued these lectures in the following term.[63] On this topic he would publish rather strongly, and the exchange with Einstein on the priority placed on central contributions to the final formulation of general relativity earned his work on this particular field of physics wide recognition (Sauer 1999; Rowe 2001; Sauer and Majer 2005, 259–276) (Table 3.4).

3.5   From the "Gastprofessur" to a New Professorship for Peter Debye

When the physician and amateur mathematician Paul Wolfskehl of Darmstadt died in 1906, he had decreed by will that 100,000 Marks of his assets, many years' worth of a professor's salary, should become a prize to be donated to the person who proves or disproves Fermat's last theorem. It was up to the Göttingen Royal Society of Sciences to form a commission for setting up procedures and deciding on the award. It assembled the physician Ernst Ehlers (one of two secretaries of the Society), Hilbert, Klein, Minkowski (after his death Landau) and Runge.[64] Obviously, Hilbert was in charge, as he signed for all reports and was identified as such[65]: all accounts of Wolfskehl's grant identify Hilbert as dealing with the decisions, especially while Klein was ill for some time from winter 1911.[66] Even before the prize competition

---

[63] Sauer and Majer (2009) with a short discussion of the context of the lectures p. 74–78. Hilbert introduced the lectures as follows: "Die Vorlesung, die ich in diesem Semester mit "Grundlagen der Physik" angezeigt habe, soll sich wesentlich mit den modernen relativistischen Ideen beschäftigen [...]" (81). Part 1 comprises e.g. §26 "Elektrodynamik auf Grund der atomistischen Hypothese" and §§27–30 "Die Miesche Theorie," the notes of part 2 of winter 1916/17 constitute a presentation of general relativity.

[64] Bekanntmachung vom 27 June 1908 (Einstein CP, Vol. 5, 502).

[65] According to Günter Frei, Hilbert was president of the prize commission, see Frei (1985, 136). Cf. also *Bericht der Wolfskehlstiftung* which were published in Nachr. GWG on a yearly basis from 1909 to 1921.; (Runge 1949, 149); Debye interview 1962, session II. (AHQP).

[66] Among the descriptions of Kleins illness are (Frei 1985) ("Klein mußte nach erneuten depressiven Anfällen im August 1911 nach Hahenklee zur Erholung und Ende November ein zweites Mal. 1912 verbrachte er dort insgesamt neun Monate. Auf 31. Dezember 1912 bat er um Entlassung vom Amt in Göttingen."), the *Mathematisch-Naturwissenschaftliche Blätter* 1912, p. 27, which reported that Weyl took over his lectures after Christmas 1911, and Tobies (1981, 86) ("So trat im November 1911 eine so schwere Erschütterung seines körperlichen Zustandes ein, daß er ein ganzes Jahr in einem Sanatorium in Hahenklee im Harz verbringen mußte.") Also (Runge 1949, 152), mentioned that Klein stayed summer and fall 1912 in a sanitarium. Although Runge acted in his place, he kept organizing educational matters. Cf. also his declining number of contributions to the *Mathematische Gesellschaft*.

was officially opened in June 1908,[67] it had already become clear that no quick winner would turn up, so that it appeared to be more reasonable to use the interest of the money to support mathematical work related to the problem.[68]

Table 3.5 summarizes the spending of this money together with some additional funds that were requested from the state guest professorships. As one can easily recognize, only a small fraction was awarded for the purpose of Wolfskehl's foun- dation. In the first year of operation less than a third of the available interest was spent for work on the Fermat problem by a mathematician in Münster.[69] No further appropriate spending was found until ten years later, when a book on the Fermat problem earned 1500 Marks in 1919.[70]

Besides its true, albeit neglected purpose, the Wolfskehl funds were, in fact, spent on guest professorships for mathematicians and theoretical physicists, but also for talks by experimental physicists and biologists or philosophers. The first Wolfskehl-funded *Gastprofessor* was Henri Poincaré in 1909. He chose to speak on integral equations and their application to physics and astronomy, in particular on tides and Hertzian waves, as well as on transfinite cardinal numbers. While not concealing the competition between Hilbert and him, he made the event at the same time a harmonious demonstration of scientific internationalism.[71] The *Lorentz-Woche* in the following year was a similarly high-profile event and stood under the title "On the development of our conceptions of the aether." According to Max Born, Lorentz gave "a good survey on the physics of that time and culminated in the derivation of Planck's radiation formula."[72] While Voigt had been exchanging letters with Lorentz for decades, in particular on his research on the Zeeman effect, and had invited him to Göttingen in 1897—and thus must have known him far better than his mathematical colleague—, it was Hilbert who introduced the guests and spoke about the relationship between physics and mathematics in the opening and closing addresses.[73]

In 1911 the German-American professor exchange program brought German-born American physicist and Nobel laureate of 1907 Albert Abraham Michelson to Göttingen. With him came a prestigious new diffraction grating of his own produc-

---

[67]The Wolfskehl-Preis was presented to the public on 27 June 1908 [...], cf. Jb. DMV (1909, pp. 111–113).

[68]This caused problems with the widow, who was against having the prize money divided. Cf. postcard from Minkowski to Hilbert 9 May 1908, in Minkowski (1973).

[69]Jb. DMV (1910, p. 175).

[70]Nachr. GWG (1919).

[71]Hilbert Papers, folder 579, Nr. 1, Bl. 1–2. Report in Jb. DMV (1909, p. 39). Poincaré lectured in German, the lectures were published as Poincaré 1910. For an account of the honored visit cf. Reid (1970, p. 120). In Hilbert's opening address on April 22, 1909, he proclaimed that, from a mathematical perspective, Germany and France could be considered a "single country" ("Die mathematischen Fäden zwischen Frankreich und Deutschland sind so mannigfaltig, und stark, wie nirgends zwischen zwei Nationen, so dass wir in mathematischer Hinsicht Deutschland und Frankreich als ein einziges Land ansehen.").

[72]Announced title "Über die Entwicklung unserer Vorstellungen von Äther" (Jb. DMV, 1910, p. 227), published as (Lorentz 1910), quote from Born (1975, 207).

[73]Cf. his notes Hilbert Papers, folder 577.

Table 3.5 Overview of the spending of Wolfskehl funds and means for *Gastprofessur*

Date	Speaker/grants for	Comments
22–29 April 1909	Poincaré	Six talks on integral equations and relativity, 2500 M.
1909	Wieferich	Mathematician in Münster, 1000 M. for papers on Fermat problem
24–29 October 1910	Lorentz	Lorentz-Woche, talks "On the developments of our conceptions of ether," discusses Planck's radiation formula
June 1911	[Michelson]	Lectures by exchange professor [German American professor exchange], no Wolfskehl money spent
1911	Zermelo	5000 M. for his works on set theory and as a grant to allow his recover from illness
July 1912	Sommerfeld	1000 M. for lectures on quantum theory in Hilbert's class (Laue's discovery was presented in *Physikalische Gesellschaft*)
21–26 April 1913	Planck, Nernst, Debye, Lorentz, Sommerfeld, von Smoluchowski (Einstein declined)	Congress, "Gaswoche" on "the kinetic theory of matter and electricity" 4800 M. spent (800 M. each)
Spring 1914	[Lorentz planned as model *Gastprofessor*]	Requested 5000 M; not granted by the ministry
Summer term 1914	Haar	Guest professor for theoretical astronomy (lectures on cosmogony); 2000 M. paid by Wolfskehl funds
Summer term 1914	Debye	Guest professor for theoretical physics; 1000 M. paid by Wolfskehl funds and 1000 M. support from the ministry
July 1914	[Bohr]	Talk and conversations, no funding
Winter 1914/15 until summer 1916	Debye	To raise professorial salary above usual state limit 2000 M. from Wolfskehl fund and 2000 from Voigt annually
Spring 1915	[talks planned]	Lecture series cancelled due to war
In summer 1915	Pohl E. Meyer	Job talks by experimental physicists for Riecke position
June 1915	Born Sommerfeld Einstein	Talk in Math. Ges. on crystal structure, 7 June talk in Math. Ges. on modern physics, 15 June six talks, one in Math. Ges. on gravitation, 29 June
Spring 1916	Smoluchowski	Lectures in mathematical physics
4–6 June 1917	Mie	Three talks on Einstein's theory of gravitation and matter, had been planned for 1916
25–29 June 1917	Hecke	Lectures on mathematics
11 December 1917	Born	Talk on liquid crystals in *Math. Ges.*
[1918]	[Ehrenfest invited]	Refused to come
14–17 May 1918	Planck	lectures on the current state of quantum physics
16–19 December 1918	Driesch	Talks on "organic causality"
1919	Bachmann	Paul Bachmann for book "Das Fermatproblem" 1500 M.
1920		?
1921		No Wolfskehl funds spent
1922	Bohr	Invited for 1921, money for *Gastprofessur* requested
[1922]	[Russell invited]	Money for *Gastprofessur* requested

tion, which was to allow Voigt better research on the Zeeman effect. As Michelson thus occupied the place of the *Gastprofessor*, nobody else was invited on Wolfskehl funding.[74] For this reason, Hilbert was able to use the available money to support Zermelo instead, who was ill and had not succeeded in finding a position.[75]

In the following year no lecture week took place for various reasons. In one rare instance, Sommerfeld received a generous salary for two talks. In 1913 a more ambitious project was realized, an international congress. Paying each of the six guest speakers 800 Marks allowed to finance the congress fully from the roughly 5000 Marks annual interest.[76] From 1914 on, Hilbert tried to change the character of the yearly events from Pentecost lecture weeks to stays for an entire summer term for guest professors. At the same time, he applied for state funding for this new format. He had probably taken Nernst as a model, who had, on the one hand, realized the influential Solvay conferences with help of the Belgian industrialist Ernest Solvay, and on the other hand, taken advantage of this event to argue for funding from the German state as well as from private sources.

In June 1913 the Göttingen Society of Sciences officially, and Hilbert in a number of separate letters, applied to the Ministry of Culture for funds to invite a leading scientist in either one branch of pure mathematics or one from a neighboring field like theoretical physics or mathematical epistemology.[77] The *Gastprofessur* would establish a special research forum for theoretical research, which should get funds"like those received by the experimental sciences to large extent, in particular by the research institutes of the Kaiser Wilhelm Society."[78] Hilbert's call for a kind of Göttingen Kaiser Wilhelm Institute for theoretical research was given weight by reference to the exceptionally modest means needed for it.[79]

The leading scientists Hilbert had in mind in 1913 were H. A. Lorentz, Ernest Rutherford and Jacques Hadamard. Lorentz was first choice for two reasons: Having resigned from university duties and turned completely to research at the age of 58, he would bring in weight and availability. In addition, "as a brilliant model" he would guarantee a good start that could be used in the following years to acquire state funding for further scientists.[80] Furthermore, it was Lorentz who had made the Solvay Conference a successful forum for quantum theory.[81] Rutherford was another

---

[74]In Born (1975, 207), Michelson is put in one line of guest professors with Poincaré and Lorentz. He found his lectures, however, less inspiring than his tennis play and family.

[75]Report in Nachr. GWG (1911, p. 125).

[76]Document 4 (see Chap. 6).

[77]Document 4 (see Chap. 6). For the correspondance on the application cf. GStA PK 76 V c, Sekt. 1, Tit. 11, Teil 9 Nr. 10, Bd. 7.

[78]Dokument 3.

[79]Document 4 (see Chap. 6).

[80]Hilbert to Krüss, 10 January 1914, Hilbert Papers, folder 494, Nr. 10, Bl. 33. In Hilbert's words: "[...] möchte ich nunmehr alle Mittel anstrengen, um H. A. Lorentz Leiden dieses Frühjahr doch hierher zu bekommen—nicht um mindestens auch deshalb um ein glänzendes Präjudiz und Vorbild eines Gastprofessors zu schaffen, das uns später bei dem Kampf um den Etat für 1915 nützlich sein soll."

[81]Cf. Barkan (1993), Schirrmacher (2012).

leading physicist who had contributed to the atomic debate, hosted Bohr when he was writing his seminal papers, although he was not fully convinced of the later theory with its quantum jumps. Hadamard, finally, had a good record in more standard mathematical fields of great relevance for physics like differential equations.

Hilbert took considerable pains to argue for his proposals. In a long letter to Hugo Krüss, a physicist working for the Berlin ministerial bureaucracy, he did not hesitate to attempt to explain the basics of the theory of relativity to the representative of the *Kultusministerium* in order to expound how important this *Gastprofessur* was, particularly for a subject that could only be solved by the joint effort of mathematicians and physicists: the atom.[82] And in a similar way, Hilbert reported to *Ministerialdirektor* Naumann about Debye's work on a later occasion.[83]

When the Ministry of Finance ultimately rejected the application, Hilbert was shocked and reiterated his claims on a reduced level.[84] Lorentz, however, did not come. The cheaper solution bought two guest professors: Alfred Haar was paid for his lecture on theoretical astronomy from the Fermat fund, while Debye, as *Gastprofessor* for theoretical physics, received the reduced state contribution as well as Fermat money in equal parts.[85]

Whether the two guest professors of summer 1914 achieved what Hilbert had promised to the ministry is hard to decide. Parts of the lectures can be reconstructed from two folders in Hilbert's papers. From Haar's lectures Hilbert kept a four-sheet manuscript on Maxwell's theory of the Saturnian ring.[86] A different folder contains manuscripts and notes of three lectures by Debye. Only the first one on electrical dipole momenta is dated, 4 May 1914. The others treated Rutherford's and Bohr's atomic models.[87] It is likely that these talks were given in preparation for Bohr's visit in Göttingen, who was en route to Munich and then for the summer holidays in Switzerland in July 1914.[88]

Due to the war, the Pentecost workshop was canceled in 1915.[89] Part of the available money now went regularly to Debye to enhance his professorial salary.

---

[82] Document 5 (see Chap. 6).

[83] Document 11 (see Chap. 6).

[84] Refusal from Ministry of Finance to the Ministry of Culture, 11 November 1913, GStA PK 76 V c, Sekt. 1, Tit. 11, Teil 9 Nr. 10, Bd. 7; Hilbert to Krüss about his "Schreck" and the need to get at least 1000 Marks from the overall 4000 needed from state funds, 10 January 1914, Hilbert Papers, folder 494, Nr. 10, Bl. 33.

[85] Document 7 (see Chap. 6). See also report by Hilbert in Nachr. GWG (1914, p. 17), which details that Haar was lecturing on cosmogony and gave also talks in the "Mathematischen Gesellschaft and the "mathematisch-physikalisches Seminar."

[86] "Maxwells Theorie des Saturnrings"; Hilbert noted on the envelope "Haar als Gastprofessor S.S. 1914." Hilbert Papers, folder 699a, Bl. 1–12.

[87] "Elektrische Dipolmomente. Vortrag gehalten am 4. Mai 1914," "Rutherfordsches Atommodell," and "Bohr'sches Atommodell." Hilbert Papers, folder 693, Nr. 1, Bl. 1–8, Nr. 2, Bl. 9, and Nr. 3, Bl. 10–15.

[88] Bohr was in Munich on 15 July 1914, cf. Eckert (1993, 53). For his impression of Göttingen see letters in Bohr CW (2019, Vol. 2).

[89] Nachr. GWG (1915, p. 14).

Funds were also spent to invite Born and Sommerfeld for single talks and Einstein for a lecture series of six talks, which spread over the various seminars at the end of the summer term, thus, in effect, creating an equivalent of the lecture weeks with a moving venue. With Einstein, Born and Sommerfeld and possibly other visitors for the talks, Göttingen demonstrated at the end of the first year of war that scientific life went on.

Another way to spend Wolfskehl money was to invite a number of physicists for the job talks that became necessary to fill the vacancy after Riecke's death, many of which probably would not have taken place. Although in some official reports these talks were counted as scientific exchange, their only purpose was to get to know the capabilities of applicants for Riecke's succession.[90]

Hilbert's growing interest in gravitation and Mie's field theory made it desirable to invite the latter for a spring workshop in 1916. This had to be postponed until 1917, when Mie was finally able to present in three lectures the problem of matter within Einstein's theory of gravitation, which were subsequently published in a trilogy (Mie 1917a, b, c). In 1916 it was Merian von Smoluchowski who gave lectures on mathematical physics.[91] 1917 actually saw two lecture series: besides Mie, also Hecke lectured on mathematical problems. And Born came, too, to deliver a talk at the *Mathematische Gesellschaft* that was also supported by Wolfskehl money, demonstrating another use of the Wolfskehl funds: maintaining good relations with former scholars.[92] While Ehrenfest refused to deliver the by then rather prestigious Wolfskehl lectures in 1918, citing as reasons the "continuous rape of Belgium" and the "sympathies with the views of my closer Russian friends,"[93] Planck lectured in this time slot instead. He had promised a summarizing report on quantum theory and suggested to invite also Sommerfeld and Paul Epstein (form Strasbourg) to this event. In his case, support from the Wolfskehl fund was planned, but in the end not realized.[94]

The fact that Hilbert also invited Hans Driesch, a theoretical biologist and neovitalist who was later known for para-psychological studies, to give "physical-philosophical lectures," especially "on organic causality," may seem rather strange at first glance. However, in the context of Hilbert's interest in extending his axiomatic thinking step by step to all sciences, Driesch appears less of an outlier; Runge and Debye were reported to have inspired Driesch considerably during his stay.[95] After

---

[90]Cf. Debye's report in *Chronik der Georg-August-Universität zu Göttingen* 1915, p. 52, and Nachr. GWG (1916, p. 13). Debye acknowledges in a letter to the ministry from 28 August 1915 that funding came from the Wolfskehl foundation "which Hilbert made available in a thankful manner to invite the candidates in question, as far as they are unknown in Göttingen, for public scientific lectures." (GStA PK 76 V a, Sekt. 6, Tit. IV, Nr. 1, Bd. XXIV, Bl. 268–269.

[91]Nachr. GWG (1917, p. 13).

[92]Nachr. GWG (1918, p. 40); Mie to Hilbert, 16 May 1917, Hilbert Papers, folder 254, Nr. 6, Bl. 9

[93]Ehrenfest to Hilbert, 11 March 1918, Hilbert Papers, folder 91, Nr. 2. He, however, also writes that he presumably would not get a passport.

[94]Hilbert to Sommerfeld, 18 February 1917, DMA, Sommerfeld Papers.

[95]Nachr. GWG (1919, p. 43). For Hilbert's turn to philosophy cf. Peckhaus (1990). On his Göttingen stay see Schnaxl to Hilbert, 27 December 1918, Hilbert Papers, folder 347.

the war, only one more grant was given, actually for a publication on Fermat's problem, before inflation destroyed this influential asset. A major resource for physics development was gone. The invitation for a guest professorship in 1921 and 1922, which was now fully dependent on state funding, went to Niels Bohr and Bertrand Russell. In this way, the renowned *Bohr-Festspiele* (Bohr festival) took place in 1922.

3.6   Networking Göttingen Nationally and Internationally

In the last couple of sections I have tried to quantify various types of investments of resources in a new physics at Göttingen. Through surveying personnel, teaching, talks and invitations to guests, a specific local picture was drawn for a process of interaction and reorganization of physics and mathematics. While the guest professors and further invitations of scholars from Germany and from abroad already exhibited a certain level of cooperation with other scientific centers, I would like to look now briefly at particular efforts to connect Göttingen with a kind of quantum discourse, that had just emerged at national and international meetings, some of them in rather select circles.

As Thomas Kuhn had recognized in his book on the early quantum theory, a broader discussion and a significant rise in quantum papers occurred in 1911, related to new attention to the field of specific heats, which in particular Arnold Sommerfeld popularized at the 1911 *Naturforscherversammlung* in Karlsruhe. This event was the main annual meeting of physicists and mathematicians alike (Kuhn 1978, 216–219). A few weeks later, a much more intimate circle gathered in Brussels for the first *Solvay Council* on "radiation theory and quanta," an exclusive event orchestrated by Walther Nernst and financed lavishly by the Belgian industrialist Ernest Solvay, gathering the top European physicists to confer on finding a solution to the quantum riddle.[96] Clearly, no quick solution was found, and the Solvay Conferences became an ongoing endeavor; the second one convening in 1914.

With these two series of national and international events that focused on quantum problems from 1911 on, a relevant context for the Göttingen efforts to connect to this broader discourse can be identified. Two Göttingen initiatives reached out in this way: the invitation of Arnold Sommerfeld in 1912 and the *Gaswoche*, a conference on the topic of kinetic theory of matter (Born 1913a).

The first attempt to get the new quantum discourse to Göttingen was to ask a participant to share his knowledge. This may be the explanation for Hilbert's invitation letter to Sommerfeld in April 1912. Here he gives various, albeit hardly conclusive reasons why no "Fermat weeks" of the usual kind should take place: On the one hand the international congress of mathematicians in England was stealing the show, on the other hand Klein was ill and could not take part. "For these reasons," he contin-

---

[96]Participants: Nernst, Planck, Rubens, Sommerfeld, Warburg, Wien; Jeans, Rutherford; Brillouin, Curie, Langevin, Perrin, Poincaré; Einstein, Hasenöhrl; Kamerlingh Onnes, Lorentz; Knudsen, cf. Instituts Solvay. Conseil de physique (1921).

ued, "I have thought of the following alternative of more modest dimension: As I am teaching in this term on principles and axioms of physics [...], what would you think about replacing me in the last two sessions, [...]? This time would fit the Göttingen lecturers and younger mathematicians and physicists presumably best, such that I could guarantee for a well-filled auditorium."[97] Furthermore, Hilbert offers 1000 Marks form the Fermat fund to finally convince him. As he politely leaves the decision about the topic to Sommerfeld, at the same time he mentions that the theory of radiation and quantum theory would be most welcome.

Hence this letter is a good example that shows a direct effort to redirect financial means meant for mathematics to fields of his own special interest, viz. radiation and quanta. It is, however, hard to relate the investment of more than a year's salary for Hilbert's physics assistant Ewald, or roughly the monthly income of a full professor at that time, with the returns of two concluding talks for a lecture course by Hilbert.

Hilbert's lectures were ultimately entitled "Theory of Radiation" (Strahlungstheorie)[98] and, in any case, it was an unusual course. This is apparent from a report by the Göttingen *Mathematischer Verein*, which not only stated that members were recommended to register for practical classes in physics at the earliest possible date, with reference to the overcrowded institutes at that time, but also pointed out the special free lectures on the foundations of physics by Hilbert, suggesting that these offered a measure of relief.[99] This is an example of Hilbert's forceful entry into physics teaching. As it was a lecture without the usual fees, in economic terms it constituted a subsidy for its field of research. As part of this free offer, highly paid Sommerfeld delivered what he had been asked to. Moreover, he used his presence to report to the *Physikalische Gesellschaft* on Laue's discovery of X-ray diffraction with crystals.[100]

A second step to get in touch with the quantum and Solvay people followed, when the *Deutsche Mathematiker-Vereinigung*, in collaboration with the physical section of the *Gesellschaft der Naturforscher und Ärzte*, organized a joint session at the Münster *Naturηforscherversammlung* in September 1912. Under the chairmanship of Sommerfeld, Hilbert the only mathematician, had announced a speech "On the foundations of the kinetic theory of gases" but finally gave a talk on "The foundation of elementary radiation theory," while Nernst reported "On the energy content of gases" and von Smoluchowski discussed "Experimentally detectable molecular

---

[97]Document 1 (see Chap. 6).

[98]The announced "Prinzipien der Physik" were postponed to the winter term, notes are entitled "Molekulartheorie der Materie."

[99]Mathematisch-naturwissenschaftliche Blätter, 1912, p. 27: "[...] vierstündige Gratisvorlesung über die Grundlagen der Physik,' zur Entlastung der überfüllten Institute." For the nearly steady growth of student numbers in mathematics and natural sciences from 1893 to 1913 by a factor of 10, see Lorey (1916, 22); the numbers of staff only roughly doubled in the same period.

[100]Mathematisch-Naturwissenschaftliche Blätter 1912, p. 187. "[Sommerfeld] im Kolleg von Herrn Geh. Rat Hilbert eine Reihe von Vorlesungen über die Quantentheorie hielt. Er referierte in der Physikalischen Gesellschaft auch über die Lauesche Entdeckung der Beugung von Röntgenstrahlen durch Kristallgitter."

phenomena that contradict ordinary thermodynamics."[101] In effect, this program anticipated the later Göttingen *Gaswoche*.

The conference "on the kinetic theory of matter and electricity," as it was officially called, was exceptionally well organized and advertised. Hilbert inquired about work in the field of the theory of gases and radiation a full year before the meeting, invitations were sent half a year in advance.[102] Several weeks before the meeting, it was the cover story in the *Naturwissenschaften*, which even published abstracts of the talks (Born 1913a; Planck 1913).

The unidentified author of the announcement in *Naturwissenschaften* was Max Born. He wrote on the aims of the congress which was presented as a direct follow-up to the Solvay meeting, that while the latter "gave the best overview on the state of quantum theory" for the Göttingen congress "the foundation of this teaching [Lehre] comes to the fore and, what is most closely related, the revision of the foundations of statistical mechanics." The general situation was characterized in the article by the claims that the phenomenological approach had lost its legitimacy, the molecular viewpoint was the truly justified one, and that one must not wonder about "claiming forces and relations for molecules, that are foreign to our macroscopic experience." "Phenomenological modesty [Begnügsamkeit]" was out, and a "striving for a deeper insight in the essence of matter is re-awoken" (Born 1913a, 297 and 299). Moreover, "it is possible to say that for the time being it is the noblest duty of physics to search for practicable assumptions about the physics of molecules."

Planck's quantum hypothesis served as prominent model and turned out to have far-reaching applications. But "is it really an expression of a fundamental new law of atom mechanics [Atommechanik] or can it perhaps be explained on the grounds of usual mechanics?" An open question for the meeting was hence, "whether the contradictions, quantum theory exhibits against our other concepts will be cleared up easily by revising the foundations of statistical mechanics," which would possibly allow classical mechanics to be maintained, or whether the quantum had to be acknowledged as something irreducibly new (Born 1913a, 298 f). In the abstract to his Göttingen contribution Debye employed rather an opposite approach, stating that statistical mechanics probably had to be "corrected" by quantum theory, stressing his firm belief in the irreducibility of this concept (Planck 1913, 140–145).

The proceedings of the Göttingen congress appeared in 1914 at the time when those of the Solvay meeting were being reviewed and the German translation published (Born 1914b; Eucken 1914). It lists the authors as "von M. *Planck*, P. Debye, W. *Nernst*, M. v. Smoluchowski, A. *Sommerfeld* und H. A. *Lorentz* mit Beiträgen von H. *Kamerlingh-Onnes* und W. H. Keesom, einem Vorwort von D. Hilbert und 7 in den Text gedruckten Figuren." Five of eight contributors to the *Gaswoche* volume (above in my italics) had participated in the Solvay congress, while some had

---

[101]Cf. report in Jb. DMV (1912, p. 158). Hilbert's contribution appeared in Nachr. GWG (1912, pp. 773–789), also in Hilbert (1935, 217–230). Interestingly, Sommerfeld in turn gave a talk in a mathematics section, cf. ibid. p. 154.

[102]Cf. e.g. Hilbert to Einstein, 30 March 1912 and Einstein to Hilbert, 4 October 1912, (Einstein CP, Vol. 5, pp. 439 and 502).

been invited to Göttingen right after this meeting (Poincaré and Lorentz). Einstein, however, who declined the invitation to avoid distraction from his work on gravitation, further shows how the scientific groups at the two conferences matched. In his preface Hilbert stresses the relevance of the papers to mathematicians. "May this collection of talks," he wrote, "especially stimulate also the mathematicians to deal with the world of thoughts, that has been created by the recent physics of matter" (Planck et al. 1914, 1).[103]

The topic of the second Solvay conference in fall 1913 might well have been chosen by the Göttingen physicists and their *Ersatz* scientist Hilbert, and to some extent Courant's and Born's mathematical selves, as it uses exactly their phrase, the structure of matter. The invitation list would have been chosen differently, however—only Voigt, still a convinced phenomenologist, was invited from Göttingen. In the company of Nernst, Eduard Grüneisen, Heinrich Rubens, Sommerfeld, Wien, Einstein, Laue, Paul Weiss, Friedrich Hasenöhrl and others, he might have been conceptually isolated. Instead of a planned full-length paper, Voigt read only a brief report on the relation between pyroelectricity and temperature. To please and to pay homage to the congregation, he suggested at one point that the quantum hypothesis might play a role in its explanation.[104]

More in the center of the program was Laue, who was to report on X-ray interference by three-dimensional crystal lattices, as well as Grüneisen, who gave a talk on the molecular theory of solids in which he refers to Einstein, to Born's and von Kármán's theory, and in particular to Debye's *Gaswoche* speech. Interestingly, there is an *Solvay-Gaswoche* interaction in both directions, as Debye in the later *Gaswoche* proceedings also refers to Grüneisen's Solvay contribution of 1913.

Due to the topics and interaction one could call the Wolfskehl congress "Solvay 1.5," since both the field (quanta and structure of matter) and the speakers overlapped (Schirrmacher 2015, 2116–19). What the Solvay trust was to the Brussels meetings, the Wolfskehl donation was to the Göttingen activities. Actually, in both cases the initial motivations for the bequests were quite different from the later results: As Wolfskehl wanted to tie his name to the centuries-old puzzle of Fermat's theorem and would never have wished to support physics research, Solvay either had the promotion of chemistry in mind or had intended to discuss his own rather ominous pet theory with the European science elite.[105]

---

[103]"Möge diese Sammlung von Vorträgen insbesondere auch die Mathematiker zur Beschäftigung mit der Gedankenwelt anregen, die von der neueren Physik der Materie geschaffen worden ist."

[104]Instituts Solvay. Conseil de physique (1921), for a concise description cf. Mehra (1975, 74–112). "Voigt gave a brief report on the relation between the pyro-electricity of certain crystals and temperature, based on research at his institute in Göttingen. He thought that the results were of interest not only for crystals but also for the quantum hypothesis." p. 88.

[105]Solvay had prepared a re-publication of his work, and supplied it to the participants of the first Conference (Solvay 1911).

3.7 Hilbert and Voigt—Resource Differences

The discussions about Hilbert's use of resources and investments in physics call for a comparison. What were the plans and actions being pursued by the physicists at the same time, and how did they invest their resources? Clearly, a comparison with Voigt is in order. However, the differences and contingencies that make such a comparison of limited significance must not be ignored; for instance, the fact that in 1912 Hilbert was 50 years old while Voigt was already 62. This difference had less to do with youth versus proficiency, than with the risk that Hilbert could be lost to another university, while this was unlikely for Voigt.

Voigt's disciple presumed to show the highest potential was Paul Drude. Russell McCormmach speculated that Drude would have become a leading quantum physicist if he had not committed suicide in 1906 (McCormmach 1982, 100f). Voigt's assistant during the prewar period, Gustav Rümelin, might also have become an interesting example for following a different path into modern physics. He completed his doctorate with Nernst at Göttingen in 1905 and worked with Rutherford on radioactivity in 1906/07. Although he later looked into a problem originating from this in his habilitation, which he finished in 1911, he never generated much publicity, having inherited a certain reserve from Voigt. Yet neither Rümelin nor Voigt's other assistant Werner Planck had a real chance to excel: Voigt found himself writing obituaries for his assistants, both of whom were killed in action in 1915 (Voigt 1915b).[106] Nonetheless, it remains obvious that both the role of the assistants and their scientific development were quite different from those of Hilbert. This was mainly due to Hilbert's ability to mobilize resources when needed, while Voigt looked back on a long record of unfulfilled wishes.

On a cognitive and psychological level, Voigt and Hilbert differed radically. When Voigt described the character and aims of his institute in 1913, he wrote

> Only that we do not engage in pioneering work on new, unsolved fields, but pursue tasks that are related to the testing and further development of the theories of already established fields and especially measure the numeric values that appear in the theories. From the outside this may seem a rather plain occupation, it is, however, nevertheless necessary and useful. This activity, as I have pointed out previously, is closely related with higher scientific education in physics.[107]

For Hilbert there was no need to care of educational needs in this way. His use of lectures and seminars to test and expand current research turned out to be very fruitful for prospective researchers, who probably did not always achieve the full spectrum of education in mathematics and physics.

---

[106]It is striking, however, that the personalities sketched here appear to be quite different from e.g. Hilbert's assistants.

[107]Voigt at "Tagung der Göttinger Vereinigung zur Förderung der ang. Physik und Math. Protokoll vom 21.–22. November 1913 zu Göttingen" GStA PK 76 V a, Sekt. 6, Tit. X, Nr. 4 Adbih., Heft VII, p. 46 f.

There are few instances where one can more or less directly compare Hilbert's and Voigt's access to resources. While Hilbert took pains to connect both to the German physical community, which he joined at the 1912 Münster meeting, and to the Solvay circle of eminent European scientists, whom he invited to Göttingen—a group which did not consider admitting him, despite the membership of the French mathematician Poincaré—, Voigt participated in the second Solvay conference. He also requested support from the Solvay Foundation, which had been established shortly after the 1911 meeting; coincidentally, his long-time friend Lorentz was dealing with the applications. Nonetheless, the language may be telling, as Voigt writes "I kindly ask you to excuse the pestering. You know *how difficult* it is for me to satisfy the demands of the many men, who work in my institute" when he asks for financial support for an apparatus for his assistant Dr. Rohn, which would cost 600–1000 Francs.[108] As he had to realize early, his chances for getting reasonable support from the ministry were dim. Instead Klein's *Göttinger Vereinigung* provided some help, although his institute was not one dedicated to applied physics and hence, strictly speaking, outside the scope. Nonetheless, he could be grateful for "repeatedly having been granted welcomed help." The last time had been two years before, when it contributed to the procurement of an "modern first-class lattice spectrometer."[109]

In this way Voigt was well equipped for optics and crystal physics, his main areas of interest. Some of the equipment he had received directly from industrialists, and a few objects were funded by the *Göttinger Vereinigung* or even foundations like Solvay's; hence, the only option was for him to pay out of his own lecture fees or salary. Ironically, part of his income was to subsidize a project for years that had been initiated by Hilbert.

---

[108] Voigt to Lorentz, 5 June 1913, AHQP/LTZ4.

[109] Cp. Voigt's statement at the meeting of the *Göttinger Vereinigung*, November 21–22, 1913, protocol p. 47, GStA PK 76 V a, Sekt. 6, Tit. X, Nr. 4 Adbih., Heft VII.

Chapter 4
The Born "Schools" in Berlin, Frankfurt and Göttingen

Traditional accounts of the history of quantum mechanics often relate events to a few leading figures and their schools, viz. the Sommerfeld school, the Born school and the Bohr school. As suggestive as this picture may be, it tends to oversimplify the more complex state of affairs, in particular when one claims that there was, more generally, a "Göttingen school" or a "Copenhagen school," since the approach only places the case of Munich in the proper school framework. In Copenhagen, we find something more like a postgraduate institute, as the activities of Bohr's institute depended heavily on constantly changing combinations of postdocs.[1] As we have seen for Göttingen, various directions of physical and mathematical, experimental and theoretical, phenomenological and reductionist efforts coexisted and competed. Göttingen physics and mathematics exhibited several smaller learning and research collectives; therefore we can find important groups of scientists that both coordinated their research and fought to extend their spheres of influence.

In the school categorization, Peter Debye was a disciple of Sommerfeld. However, he grew more and more independent of his Munich teacher in Göttingen, as he did elsewhere later on.[2] Similarly, Max Born would turn to new directions at Berlin and Frankfurt after leaving Göttingen in 1915. However, to a greater extent than Debye, he managed to remain part of a group of researchers who followed similar lines of investigation, and these collaborations also exhibited very different qualities and forms, which were the result of circumstances much different from Debye's and Hilbert's comparatively calm and peaceable Göttingen.

---

[1] See Kojevnikov (forthcoming).

[2] A declaration of independence from his teacher is apparent from an incident in 1916 when Debye angered Sommerfeld by publishing in his domain (Eckert 2013, 52f.).

© The Author(s), under exclusive license to Springer Nature Switzerland AG 2019   61
A. Schirrmacher, *Establishing Quantum Physics in Göttingen*,
SpringerBriefs in History of Science and Technology,
https://doi.org/10.1007/978-3-030-22727-2_4

Born took up the extraordinary professorship of theoretical physics in Berlin, which Planck had made possible not without some complications.[3] In his memoirs he later wrote a vivid account of his military service in a special group of scientists that was trying to improve German sound ranging efforts. Apart from much anecdotal fare, two main points are apparent. First, Born did not interact much with Berlin colleagues and students in the field of physics; instead he maintained collaborations with friends and colleagues from Göttingen and Breslau, where he also had studied. Secondly, the war experience and his scientific war effort brought him into contact with experimental and applied physics.[4] Hence, like Debye, he turned to experimentation during the war and in its immediate aftermath, although for very different reasons. In any case, when he returned to Göttingen in spring 1921, Born had become a physicist who was also engaged in measurements and empirical research, and he even brought some experimental physics equipment as well as students of experimental physics to Göttingen. Still, he was happy not to be the "full" physicist, which still loomed large as the ultimate goal of every physics scholar and which Debye had so conspicuously represented by jumping more or less effortlessly from theory to experiment.

4.1 The Berlin and Frankfurt Outstations: Born with Madelung, Landé, Stern, etc.

Born's time in Berlin was dominated not by his university teaching and research, but by work—military and scientific—in a special army unit. As for many scientists of his generation it turned out necessary to adjust to the conditions of war and to explore ideas on mobilizing science for its efforts. In the same way as Richard Courant, who had pulled together his Göttingen colleagues to help develop new telephonic equipment after witnessing the inadequate gear of the young soldiers on the battlefield first hand, Richard Ladenburg realized that he would be more effective in the German war effort if he could create a way to improve shelling accuracy rather than become cannon-fodder himself. He convinced his military superiors to establish an office for scientific measurement within the *Artillerie-Prüfungskommission*, a department for testing army equipment. This group became a larger project for sound ranging, with the aim of improving technologies to locate enemy heavy artillery (Born 1975, 238; Froben 1972, 22–25).

Clearly, this effort had a double rationale, and the longer and deadlier the war became, the more Ladenburg tried to use his laboratory well off the battlefield to

---

[3] Planck withdrew the invitation to Born after Laue had claimed his interest and priority with respect to the Berlin position; however, as Born ultimately failed to convince the ministry of this rather informal bargaining, he was not able to finish his first lecture course—not only were his students called to the front, he himself decided to serve his country before the term had ended (Born 1975, 164).

[4] For the context of Born's group in World War I and the relation of war science and basic research cf. Schirrmacher (2009, 168–172).

save scientific talent. Born, an old friend of his, had initially joined a radio brigade and he became the first Ladenburg requested for his group. At the time of the outbreak of war, Alfred Landé was Hilbert's physics assistant and immediately volunteered as a paramedic. Erwin Madelung also came from Göttingen, were he had gradually developed from an applied physicist who did his dissertation with Hermann Simon, to an experimental physicist who had worked on the atomism of solids as Riecke's assistant, to a theoretician of crystal lattices, a field where he met and inspired Born.[5] These Göttingers, as well as Ladenburg's Breslau colleagues like Fritz Reiche and Erich Waetzmann, could be saved by requesting them for the sound ranging group[6]; for others like Born's student Heinrich Herkner the order became effective when it was too late. A further part of the group consisted of experimental psychologists since most devices ultimately relied on a listener who had to be taken into the equation.[7]

The group of physicists that gathered here is of interest for the history of quantum physics in at least two respects. First, its members were predominantly early quantum physicists who could be associated with two centers, Göttingen and Breslau (rather than Berlin). Reiche prominently published on quantum theory before and after the war; his habilitation work was on spectral lines (Reiche 1913a, b, 1921). Ladenburg himself had coauthored publications with Born and Reiche, and would contribute to the solution of the dispersion puzzles of quantum theory after the war, while his colleague Waetzmann was one of the few who had actually specialized in acoustics, a field pertinent to the war problem (Born and Ladenburg 1911; Reiche and Ladenburg 1912; Ladenburg 1921). And secondly, when a routine had been established after some hectic periods in the beginning, there were "hours without serious work," which were eventually filled with scientific discussions and even research work. In this way there existed an atomic physics group in military service during the war, which was more productive in this field than those colleagues who remained at the University of Berlin.

"As soon as I had settled down in the A.P.K.," Max Born recalled, "I took up some scientific work" (Born 1978, 179). His book on the dynamics of crystal lattices had just appeared in 1915, and he extended his interest to a theory of optical activity, which was important for chemists to distinguish asymmetric (chiral) molecules and liquid crystals. This field was the perfect bridge between Hilbert's reductionist atomism and Sommerfeld's quantum rules. As Born explained in his book, it was Hilbert's insight that the route from the "enormous number of equations of single processes in molecular physics" to the "few phenomenological laws" was manageable, since "the phenomenological laws appear to be the conditions under which the equations for the molecular processes are solvable." This method of reductionism, Born added,

---

[5] Voigt, however, had dismissed his work on the atomism of solids, cf. Sommerfeld to Wien, 1 June 1916 DMA Sommerfeld papers 10.

[6] Whether Otto Stern also joined the group late in the war remains unclear. This is claimed in Segre (1973, 219), but there is no mention by Born, who remembers first becoming acquainted with Stern in Frankfurt after the war. Cf. for others like Born's student Heinrich Herkner, for whom the order became effective after it was too late (Born 1918b, 1978, 190).

[7] Among these were Max Wertheimer and Erich Hornbostel, who worked on the psychology of listening (Hornbostel and Wertheimer 1920; cf. also Hoffmann 1994).

"reveals its power in the most beautiful way for the explanation of optical activity" (Born 1915, 13).[8] In the same way, this field also integrated the new quantum theory of Sommerfeld, which extended Bohr's theory and allowed the rules of quantum theory to be applied to atoms of higher order like oxygen and nitrogen (Sommerfeld 1916a; Born 1918a, c). In one of his next papers Born evaluated experiments with infrared radiation in order to verify Sommerfeld's ring atom models, which he considered more or less adequate. This led him to the new idea to theoretically construct crystals out of Bohr-Sommerfeld atoms (Sommerfeld 1917; Born 1918a, 1978, 180).

This line of research interest, which Born was able to investigate after "office hours" at home, since his military duties evolved into practically an eight-hour-a-day white-collar job (aside from some trips to the front), was soon extended by in-office scientific collaboration. "At this time Madelung used to sit opposite me at a big desk with many drawers," Born recalled, "Each of us emptied a drawer of all military papers and filled it with scientific books and notes." Naturally, as both Born and Madelung were interested in crystals they started to collaborate on this topic (Born 1978, 181). This work produced what later became known as the Madelung constant, and the papers of Born and Landé in which they tried to calculate the compressibility of crystals, which had yielded some values experimentally; but was explained only theoretically by phenomenological theories, by now applying first principles of atomic constitution, thus realizing Hilbert's program in physics.[9]

In the course of these works by a scientific team in military disguise[10] Born arrived at a major conclusion: the elastic properties of crystals turn out to be completely inaccurate when deduced from Bohr-Sommerfeld ring atoms. Therefore, the electrons must be distributed three-dimensionally in the atoms within the crystal. The simple flat model thus failed in this case,[11] and the problem of how electrons could move in orbits that fill three-dimensional space foreshadowed the concept of electron clouds.

In Born's recollection this is the precise starting point of a program that eventually led to the establishment of quantum mechanics (Born and Landé 1918b; Born 1978, 183). His contemporary reaction, however, was slightly different. In a letter to Hilbert, Born described the events of the German revolution that he witnessed in Berlin in

---

[8]This point was also stressed by Sommerfeld (1915, 669f.), who was pointing out Hilbert's influence to the readers of his review. "Der Verf. hat sich mit den mathematischen Gedanken erfüllt, die Hilbert in der Theorie der Integralgleichungen [...] geschaffen und in seiner Darstellung der Gastheorie zur Anwendung gebracht hat. Die beobachtbaren Gesetze erscheinen hier als die Bedingungen dafür, daß die Gleichungen für die Molekularvorgänge mit ihrer ungeheuren Zahl von Unbekannten auflösbar sind. Es ist dem Verf. gelungen, diese Methode auf die Gittertheorie zu übertragen. In diesem Sinne bildet die Widmung des Buches an David Hilbert einen sinnvollen Schmuck der Theorie."

[9]Madelung (1918), Born and Landé (1918c), Born and Landé (1918a), Born (1978, 183).

[10]It should be stressed that the work on sound ranging did go on at the same time, as it also posed many challenges. The historical scholarship here is still not sufficient and the historian faces the problem that most (German) scientists did not report on their military efforts in the same detail as they did on scientific work. Cf. Schirrmacher (2009, 2016).

[11]Strictly speaking, Sommerfeld's extension of the Bohr model had already given rise to the expectation that for many-electron atoms "space quantization" would require the now elliptic orbits to lie in different planes, cf. Eckert (2013, 33).

November 1918 and about which the newspapers had not given the full picture; as it appeared to Born, it was not yet clear whether things would tend to the better. Another revolution, so-to-speak, had, however, turned out unequivocally, that is, he had established in his calculations of compressibility that atoms in the form of rings cannot hold true. They turned out to be much too soft, and "one can infer from measurement with great precision [...] that every single atom has cubic symmetry."[12] It was Landé who would excessively exploit the theory of cubic atoms before turning to the Zeeman effect in order to distance his work submitted for habilitation from that of his superior Born, but this happened later when both were working at the University of Frankfurt.[13]

Also during the times of political revolution, Born accidentally became involved in a joint research project with Fritz Haber at the Dahlem Kaiser Wilhelm Institute, when he visited James Franck in late 1918. Haber, who had entertained some structural theories himself, was intrigued to learn of Born's ideas to calculate chemical energies from physical properties of lattices and together they succeed in spring 1919 to develop a scheme that would relate various chemical and physical lattice energies in the process of the formation of an ionic compound into what is now known as the Born-Haber cycle (Born 1919b).

Without doubt, Berlin was much more strongly affected by war and revolution than Göttingen or Frankfurt. This extended to scientific life at the universities as well. In Berlin and particularly at the physics institute, university life was taking place in the middle of the political center. It is known that Born was able to speak at the Prussian Academy of Science at least twice during his war duty and that he found time to cultivate his friendship with Einstein at this time.[14] For many reasons, however, it turns out to be particularly difficult to reconstruct the teaching of physics and its impact on the student body in a special field like atomic and quantum physics. It is still instructive to consider the relevant lectures, at least, as they had been announced in the university catalog, although they often did not take place (see Table 4.1).

In the last years before the outbreak of the Great War, Walther Nernst was lecturing on "atomistics" to a general audience of science students, while younger *Privatdozenten* like disciples of his and Rubens, Arnold Eucken and James Franck, dealt with theoretical and experimental foundations of quantum and atomic physics. During the war, lectures on atomic and quantum physics were announced only by members of two groups: those unaffected by war like Einstein, due to his Swiss nationality, and Max Weinstein (1852–1918), formally still a *Privatdozent* for geography and physics, who, however, sacrificed his work to topics of natural philosophy,

---

[12]Born to Hilbert, 14 November 1918, Hilbert Papers, folder 40A, Nr. 18.

[13]From Landé's many publications on the cubic atom cf. e.g. Landé (1920). For Landé's situation in Frankfurt and the context of his work on the Zeeman effect, see the detailed account by Forman (1970a), esp. p. 165, where Forman explains his change of topic as "a response to the exigencies of German academic careers." On Born's advice to pursue a research field with some distance to Born's work, cp. Born to Landé, 27 January 1919, SPKB Landé papers, box 1, folder 5. "[...]; nachdem Sie so lange mein wissenschaftlicher Compagnon waren, empfiehlt es sich, daß Ihre Habilitationsschrift deutlich und sichtbar ohne meine Mitwirkung entsteht."

[14]30 November 1917 and 28 June 1918.

Table 4.1 Announced lectures on atomic and quantum physics at the University of Berlin, 1913–1920

Winter 1912/13	Nernst	Neuere Atomistik (für Studierende aller Naturwissenschaften)
Winter 1913/14	Westphal	Die Gesetze der Wärmestrahlung und das ultrarote Spektrum
	Franck	Experimentelle Grundlagen der Atomistik
Summer 1914	Eucken	Die Quantentheorie
Winter 1914/15	Born	Atommodelle
	Reiche	Neuere Probleme der theoretischen Physik
Winter 1915/16	Nernst	Neuere Atomistik (für Studierende aller Naturwissenschaften)
	Born	Molekulartheorie der Kristalle
Summer 1916	Born	Kinetische Theorie fester Körper
	Reiche	Kinetische Theorie der Gase
Winter 1916/17	Weinstein	Physik der Atome, Moleküle, Elektronen und Quanten
Summer 1917	Reiche	Ausgewählte Kapitel der Quantentheorie
Winter 1917/18	Einstein	Statistische Mechanik und Quantentheorie
	Reiche	Moderne Probleme der Atomdynamik auf quantentheoretischer Grundlage
Summer 1918	Born	Atomistik
Summer 1919	Franck	Experimentelle Grundlagen der Quantentheorie
Summer 1920	Franck	Experimentelle Grundlagen der Atomistik und der kinetischen Gastheorie
	Franck	Atomtheorie
	Reiche	Quantentheorie

and was simply too old to serve the Fatherland at the time; and those at the *Artillerie-Prüfungskommission*, viz. Born and Reiche. Moreover, even when hostilities had ended, it was the former assistants of Planck and Rubens, Reiche and Franck, who would teach the experimental foundations and the theories of atoms and quanta, with Born already in office at Frankfurt.

The Prussian ministerial bureaucracy of the Kaiserreich was famous for its well-planned use of academic personnel and its workings became known as the Althoff system (vom Brocke 1980). Even a year before Friedrich Althoff's death in 1908, this monolithic authority governing German professors had been distributed among four successors. War, defeat and revolution had affected this system and created more leeway for those trying to steer a new course for higher education in a democratic society in the early postwar years. Both of Max Born's changes of universities profited from this situation. In 1919 the privately devised plan of switching the positions between Born and von Laue in Frankfurt and Berlin did not meet with any opposition (as had been the case four years earlier, when Laue tried to claim priority over Born for the Berlin position). In 1921 the ministry would again show much flexibility to meet Born's demand for an additional post for Franck.

In December 1918, the Dean of the Frankfurt faculty asked von Laue to provide an assessment of Born. Given the fact that the latter had some interest in the matter of appraising Born, he nevertheless provided an elucidating account of Born's strengths and interests, which he describes to cover three different areas. First, Born had demonstrated his authority in the field of general relativity and had even defended the theory successfully against attacks as from Ernst Gehrcke. Secondly, his publications on quantum theory had proved his "ability to identify modern scientific problems and the means for their solution on his own." The theory on the specific heats of solids that he had established together with Theodore von Kármán, although it had found a competitor in Debye's different treatment, had to be judged superior, "since it accounts for the constitution of the body out of atoms. It proves the good physical eye of Born that he had founded his considerations on the arrangement of atoms in space lattices, at a time when the space lattice theory of crystals had not yet been demonstrated by X-ray interference experiments," i.e., Laue's.[15] And, thirdly, Born would do research in a field Laue called "electron theory," by which he was referring to Born's recent works on optical activity and liquid crystals. While Laue did not mention that this field also had a quantum background related to the scope opened up by Sommerfeld's extensions of Bohr's atom, he did stress "that already a number of younger researchers (Stumpf, Landé) had joined this research and Born himself was planning experimental investigations of these matters."[16] Born's intentions to turn to experiment had thus materialized and became known in the physics community.

When Born had completed negotiations with Frankfurt, he had not only become a full professor for the first time, he also had "achieved everything [he wanted]: 3,000 M budget for my institute, an assistant and a mechanic," as he wrote Landé; at the same time, he noted, that "I necessarily need an assistant, who is not only able to do experiments with prepared equipment but also can create new apparatus. Therefore, I won't be able to give this position to you; we have to make sure that we find funds for a theoretical assistant later."[17] The theoretical, or even mathematical physicist, as which Born had widely been recognized, was now becoming a "full" physicist who naturally dealt with experiments, and he valued his experimental assistant, whom he still had to find, more highly than a theoretical one, who was standing by—in a similar way as Hilbert had temporarily transformed from a pure mathematician into a theoretical physicist and had once treated his mathematics and physics assistants.

Otto Stern would become one of the new partners in this field for whom Laue had already arranged a lectureship; another was Walther Gerlach, the assistant to Richard Wachsmuth, director of the physics institute in Frankfurt and founding rector of the University of Frankfurt, which was established in 1914. Gerlach quickly felt more attracted by Born's physics than by the activities of Wachsmuth, who remained a classical physicist in all his topics and did make no contact with relativity nor

---

[15]Laue to Curator, 10 December 1918, UA-Fra Main personnel file Born, p. 1–3.

[16]Ibid.

[17]Born to Landé, 27 January 1919, Born to Landé, 27 January 1919, SPKB Landé papers, box 1, folder 5.

quantum theory and its experimental effects. With Stern, who came from Silesia like Born, who had studied in Breslau like him and had also worked with Einstein, Born quickly became close friends.[18] Landé, who earned his living as a music teacher at a reform school near Frankfurt immediately after the war, joined Born even without a regular assistant's position, as Born had promised to facilitate his habilitation here.[19] Landé's "independent" work was on the "quantum theory of the helium spectrum," (and the Zeeman effect); as such, he was probably the first of the quantum theorists to study astronomical books like Charlier's on celestial mechanics in order to apply perturbation theory to atomic structure (Landé 1920).[20]

The impression that Born had brought quantum physics to Frankfurt is quickly confirmed by the teaching activities.[21] While Born extended his attention to Wachsmuth's research through a joint colloquium, no lectures on modern themes would come from his older colleague. Born and Stern introduced a modern curriculum right from the start with courses on quantum theory, molecular theory, a joint seminar on quantum theory, which was later joined by Landé as well, and lectures on relativity and "atomistics" for a broader audience. From the classical fields, mechanics and the theory of heat dominated (see Table 4.2).[22] After Born had left for Göttingen, Gerlach and Madelung were replacing him in part, the former in the joint colloquium with Wachsmuth and the latter in the seminar with Stern and Landé.[23]

Parallels to the teachings at Göttingen at the same time are striking, both concerning the topics Born (and Franck) announced in Berlin and Born and Stern lectured in Frankfurt, and with respect to the formats and cooperation employed. Quantum and atomic physics courses had entered the regular curriculum, as had molecular theory. The Göttingen seminar on the structure of matter run by Hilbert and Debye found its parallel in the seminars on quantum theory (or on problems of modern physics), which were held jointly by Born, Stern and Landé. Like Debye, who held a joint physics colloquium with Voigt from summer 1916 on, when he officially became a professor of experimental physics (later also joined by Pohl), in Frankfurt Born and Wachsmuth met for physics colloquia fortnightly.

There were also important differences between Frankfurt and Göttingen, in particular when it came to the involvement of mathematicians who did not, or at least

---

[18]Their common Jewish background may not have played a role at this stage.

[19]However, it took half a year before Born could write Landé that Wachsmuth and the mathematician Schönfliess had no objections, Born to Landé, 6 June 1919, SPKB Landé papers, box 1, folder 5. His thesis was accepted in Okt. 1919, Wachsmuth to Curator, 28 October 1919, UAG Main Personal File Landé, folder 1.

[20]Landé interview 1962, AHQP, p. 4. This time however, Sommerfeld's private assistant Adolf Kratzer may also have worked on perturbation theory, cf. Kratzer (1920).

[21]During Laue's tenure only one lecture course on quantum theory is known during summer term 1917, just merely one hour long and probably meant for a general student audience. Vorlesungsverzeichnis Frankfurt. In winter 1917/18 there was also a course on relativity (two hours long).

[22]For Stern's (intended) teaching as advertised in the university catalog during the war, including an "Introduction to Quantum Theory" in summer 1916, cf. Schmidt-Böcking and Reich (2011, 39).

[23]In summer 1921 Gerlach also lectured on "Atom- und Molekülbau (für Hörer alle Fakultäten)" as well as on "Höhere Experimentalphysik."

Table 4.2 Lectures by Born, Stern and Gerlach as well as joint seminars with Wachsmuth, and Landé at the University of Frankfurt, 1919–1921

(Wachsmuth)	Born	Stern	(Landé)
Spring 1919[#]	Einf. theoretische Physik	Einf. Thermodynamik	
Summer 1919	Analytische Mechanik Quantentheorie Einf. theoretische Physik* Physikalisches Kolloquium	Molekulartheorie I: Kinetische Gastheorie	
Winter 1919/20	Mechanik II Übungen zur Mechanik II Seminar über Quantentheorie Physikalisches Kolloquium	Molekulartheorie II: Statistische Mechanik Thermodynamik	
Spring 1920	Vektoranalysis Seminar über Quantentheorie		
Summer 1920	Theorie der Wärme Relativitätstheorie in elementarer Darstellung[(1)] Physikalisches Kolloquium Seminar über Probleme der modernen Physik	Analytische Mechanik + Übungen	Theorie der elektrischen Schwingungen
Winter 1920/21	Theorie der Elektrizität Atomistik[(1)] Physikalisches Kolloquium Seminar über Probleme der modernen Physik	Mechanik der Continua	Einf. in d. math. Behandlung der Naturwiss.

[#] Special term for war veterans ("Zwischensemester für Kriegsteilnehmer"), Feb. 3 to Apr. 16, 1919
*Particularly for war veterans, free
(1) one hour long

not visibly, collaborate with the physicists. If one digs a bit deeper, however, some relationships were still present on a much smaller scale. Ernst Hellinger, who got a extraordinary professorship in 1914 when the university was founded, turns out to have become Born's "tame mathematician" after 1919, as he resided in the attic of Born's house, and, more than that, he was an old friend from school in Breslau and his student days at Göttingen, where he was a student assistant to Hilbert and the like (Born 1978, 190). His expertise, in particular on quadratic forms and infinite-dimensional matrices, would become a crucial resource only later, however, when Born was working out matrix mechanics.

It was no easy task for Born to find an experimental assistant with the abilities described in his letter to Landé.[24] After Göttingen graduate Hans Kost, who did his doctorate with Riecke, Breslau graduate Hedwig Cohn, who received her degree from

---

[24]Even before Born received the final confirmation for the Frankfurt position he tried to persuade Ludwig Prandtl to do experiments on compressibility of halides like LiF, LiCl, LiBr or LiJ for him. DLR Archive, GOAR 3664, Born to Prandtl, 2 January 1919.

Lummer, and Reinhold Fürth from Prague, a student of Philipp Franck, had declined, Born wrote to Stefan Meyer in Vienna for help. Eventually, the position was filled by Elisabeth Bormann, who had just completed her doctoral examination. She belonged to a new generation of female physicists who were given the opportunity to enter a scientific field mainly due to the war deployment of their male fellow students.[25] Having attended various practical classes with Franz Exner and Stefan Meyer (one for beginners, one on radioactivity and also an advanced one) and having written her thesis "on the experimental methodology of decay fluctuations" she had a thorough background in experimental work.[26]

With Stern and Bormann as key collaborators, plus the mechanic Adolf Schmidt and Gerlach visiting, in 1919 Born had established a strong experimental group, while Landé remained the only pure theoretician at his department, with whom Born did not publish any further joint work from this point on. Four papers appeared during the Frankfurt years co-authored with Bormann, two with Gerlach and one with Stern. Their research work corresponded to three areas of physical problems which Born tried to push forward. First, his November revolution of the failure of plane atomic models led to experimental work investigating real lattice structures of simple substances, in particular, zinc blende (ZnS).[27] Second, Born supported Stern's idea to realize molecular rays, which allowed for various experimental investigations of atomic and molecular physics.[28] Like Born, Stern, too, had spent most of his time on theory, but now he was fully committed to examining fundamental theoretical concepts through experiments.

Stern initially tried to verify Maxwell's velocity distribution and to measure mean velocities, and he entertained plans for a molecule spectrograph. Born and Bormann adopted his technology and hoped to detect cross sections of molecule collisions in order to determine, for example, the diameter of the silver atom. These experiments occupied Born to a great extent, as his wife Hedwig confided to Einstein: "Max is very hard-working, his experiments [...] finally work and he sits in the institute until 8 p.m. and performs measurements."[29] Born, of course, did not convert fully to experimental physics and his immersion did not get as far as Debye's, who took over the experimental physics position in Göttingen, but he made every effort to count as a "full" physicist who could combine theory and experiment for their mutual enrichment.

With the improved mastery of molecular rays, of silver in particular, Stern devised a test of a theory Sommerfeld had put forward to account for atomic states in his

---

[25] Born to Landé, 12 February 1919, SPKB Landé papers, box 1, folder 5; Born to Meyer, 31 March 1919 and 28 April 1919.

[26] Lebenslauf, 19 July 1919, UA-Fra Personal-Hauptakte Bormann, Bl. 3–4.

[27] The relevant papers from Born are (Born and Landé 1918b; Born 1919a; Born and Bormann 1919a, b; Born 1921a).

[28] The idea came from Hans Kallmann and Fritz Reiche in Berlin who tried to demonstrate that individual polar molecules carry an electric dipole moment (Kallmann and Reiche 1921) as acknowledged in Stern (1921, 250).

[29] Born (1920b), Stern (1920). Born to Sommerfeld, 5 March 1920, (Sommerfeld 2004, 74–75, Born 1978, 195). Hedwig Born to Einstein, 31 July 1920, (Born and Einstein 1969, 55).

quantum theory of atoms and which implied "space quantization," i.e. a peculiar behavior in the presence of inhomogeneous magnetic fields, which, if true, would result in splitting up a molecular ray into components. After Stern had checked the feasibility of an experimental test in a theoretical calculation, he was, with the help of Walther Gerlach and his experience with vacuum technology, eventually able to prove that the effect was true (Stern 1921; Stern and Gerlach 1922). At the time of this fully successful Stern-Gerlach experiment, however, Born had already moved to Göttingen.

Born also collaborated with Stern and Gerlach on various other topic related to gases and crystals (Born and Stern 1919; Born and Gerlach 1921a, b). Furthermore, all his students were doing experiments, a fact Born admitted to Felix Klein in Göttingen: "there are a number of doctoral students here engaged in experimental work, which I had suggested and whom I have to supervise. Strangely enough, I don't have any theoretical doctoral student; as it seems, people avoid this direction, at least here in Frankfurt."[30] However, it does not seem far-fetched to conjecture that not only Born, but also the students experienced a certain war effect, which, at least for some years pushed experimentation, although on a larger scale the rise of theory would soon reclaim ground.

Besides this research work Born also published more general surveys. On the advances and aims of the quantum theoretical treatment of the atom he had contributed three articles to the journal *Die Naturwissenschaften*, which later were combined into a book, and he published a popular introduction to the theory of relativity (Born 1920a, 1921b). Actually, the Stern-Gerlach experiment could have hardly been financed at Frankfurt without relativity, as Born had used the public curiosity about this theory after spectacular reports of the corroborating eclipse expeditions in order to raise funds needed for experimental equipment. As he confided to Sommerfeld: "The talks, which I gave in January raised 6.000 M for my institute. With this money I really got my institute going. Stern's deflection experiment has finally succeeded nicely."[31]

The recent foundation of the university generally necessitated reliance on special funds. On the one hand, there was still the *Physikalische Verein*, which had been the nucleus for the physics institute and which continued to provide support,[32] and, on the other hand, there were philanthropists of the wealthy city of finance, which could provide relief even—or rather particularly so—in times of inflation and financial crisis. The Moritz and Katharina Oppenheim foundation had funded the university chair in theoretical physics, which after Laue now Born occupied.[33] As Born's audacious

---

[30]Born to Klein, 21 November 1920, UAG Klein papers, folder 5D, p. 79. "Ich habe viel zu tun, weil mir sehr viel daran liegt, eine experimentelle Arbeit vor meinem Fortgehen fertig zu stellen. [...]; es sind hier mehrere Doktoranden mit Experimenatlarbeiten beschäftigt, die ich angeregt habe und überwachen muss. Merkwürdiger Weise habe ich aber keinen theoretischen Doktoranden; es schein, dass die Leute diesen Weg scheuen, wenigstens hier in Frankfurt."

[31]Born to Sommerfeld, 5 March 1920, (Sommerfeld 2004, 74).

[32]For the role of the Verein in establishing and supporting the physics institutes, see Fricke (1974).

[33]For details see papers of UA-Fra, Katharina und Moritz Oppenheimsche Universitätsstiftung, for the many further foundations cp. Lustiger (1994).

application for funds at the Kaiser Wilhelm Institute is well documented (Castagnetti and Goenner 2004), where he essentially asked for funding to set up an entire laboratory, he even spelled out a much wider experimental program, including experimental investigations about the predictions of theories of mixed crystals and metals. Despite the "large sum" he had requested, Einstein let Born's wife know that he was optimistic to raise the funds sooner or later.[34] While Debye did not take advantage of the equipment he requested from Einstein's institute, since he left Göttingen before it arrived, Born apparently managed to get funds from other sources more quickly.

To sum up, not only Göttingen with Debye but also Frankfurt with Born, Stern and Gerlach became a key place of quantum research in the war and immediate postwar years. With the reshuffling of positions in the early 1920s, some weights shifted, for example from Frankfurt to Hamburg, where Stern and Pauli pursued strong, new research fields. Concerning history and resources, Göttingen and Frankfurt were quite different: tradition versus recent innovative foundation, private funds versus state funding, rural (academic) seclusion versus urban (economic) bustle, etc. The apparent consensus about the leading fields and figures of scientific research and its overall importance, however, proves the extent to which science had attained general, even national value at a critical time in German history when it had to symbolically compensate for lacking standing in economic and military strength after a lost war.

4.2   Born's Grand Plan for Göttingen

How could the Göttingen faculty make up for Debye? In their first meeting to draw up a list of candidates, they agreed that there was but one chance to heal the debacle stylized so strongly in national terms: Arnold Sommerfeld. Although further names appeared in the minutes, Born and Mie, it was agreed to approach Sommerfeld first.[35] Göttingen chemistry Nobel Prize winner Otto Wallach, who also happened to be a distant cousin of the renowned Munich physicist, had the task of asking Sommerfeld. "Understandably enough, the wish exists to compensate for the loss, we have sustained by Debye's leaving", Wallach wrote, and he reckoned that he would "probably be completely out of reach." However, if he were willing to come, he would not only be the first name submitted to the government, he would be the only one, and he would be "greeted with the greatest rejoicing." Otherwise, Göttingen would resign itself to Sommerfeld's decision. The only incentive they had to offer was a free choice in naming an extraordinary professor for his support, which he "presumably" would get. In case of refusal, Wallach asked him for his advice on a suitable successor of Debye.[36]

As in the very same meeting a commission was put in charge of a "general reorganization" of the physics professorships—there were three of them to be recast:

---

[34] See Castagnetti and Goenner (2004).

[35] Faculty minutes, 21 February 1920, UAG Phil. Fak. II PH 36 e I.

[36] Document 17 (see Chap. 6).

Debye's, Simon's and Voigt's—and it was apparently at this stage that the solution first appeared that an appropriate successor for Debye could be found only if a second professorship were added, which the desired candidate might fill to his liking. Sommerfeld did not decline, but the faculty sensed both that the chances of his accepting were slim and that it would take too long. As in the first session, Hilbert attended the meeting, and this time Debye also became involved.[37] Again, discussions about Born and Mie emerged, as well as about Wilhelm Lenz, Madelung, von Kármán and Erwin Schrödinger. Probably, it was Debye who had pushed the last of these, as he was asked to obtain opinions on Schrödinger from Planck, Einstein, von Laue and Sommerfeld.[38] Ten days later, letters to the ministry were drafted, one for the Simon succession and one for that of Debye. Now Born was named first, with Madelung and Lenz second and third. In fact, Debye, Runge and Wallach had composed the letter for the dean to send to the ministry.[39] In this way, *Regierungsrat* Erich Wende, senior executive officer in the Ministry of Culture, actually received two long letters concerning Debye's succession and the reorganization of Göttingen physics in the spring of 1920, the other being Born's about erecting what his colleagues from the Göttingen faculty had called a *Pflanzstätte*, a nursery, for mathematical and physical sciences, albeit according to his views.[40]

First of all, Debye, Runge and Wallach told Wende that only Sommerfeld might offer an adequate substitute for what Debye had established. Trying to win him, however, was futile. Instead of Mie, who might also have been an obvious candidate, a younger scientist should be preferred. Therefore, Born was the more or less natural choice, with Madelung, who initially came from the field of technical physics and who had just taken over an ordinary professorship in Kiel, and Lenz, a longtime assistant of Sommerfeld, who had not yet published much, additional candidates. Born, the authors wrote, had excelled early in those fields of theoretical physics "for which a very thorough understanding of mathematical means is required," such as relativity with its application to moving electricity. After 1912 he had shifted towards employing his "sharp eye for the needs of physics" as in the field of specific heats where the mathematical problems appear only as secondary issues. In molecular theory, which became his major interest, he treated crystal structure, optical activity,

---

[37]Hilbert's influence on the decisions remains unclear; however he was also asked to join the commission for the succession of Simon, as a connection to Born's call, might arise. Faculty minutes, 24 June 1920, UAG Phil. Fak. II PH 36 e.

[38]Faculty minutes, 3 March 1920, UAG Phil. Fak. II PH 36 e I.

[39]Faculty minutes, 13 March 1920, UAG Phil. Fak. II PH 36 e I. The draft was modified by corrections in more than one hand, one of which appears to be Hilbert's, faculty minutes, Phil. Fak. II Ph. Nr. 36 d, XV.

[40]Draft Debye, Runge and Wallach, before 22 March 1920, faculty minutes, Phil. Fak. II Ph. Nr. 36 d, XV. The original reads: "Wir haben den dringenden Wunsch, dass die erledigte Professur durch einen theoretischen Physiker ersten Ranges besetzt wird. Das ist seit Jahrzehnten hier der Fall gewesen, und die so erfreuliche Entwicklung der mathematisch-physikalischen Fächer an unserer Hochschule, die dadurch eine Pflanzstätte für diese Wissenschaften geworden ist, wurde nur dadurch ermöglicht [...]"—The term *Pflanzstätte* is usually ascribed to Sommerfeld, who employed it in an autobiographical sketch written in 1919, which was, however, published only much later, (Eckert 1993, 38).

anisotropic liquids and the diffraction of light in gases. Steadily progressing along this line of research, which suggested, for example, that elastic forces in crystals were of electrical nature, he eventually opened up many directions for experimental research.[41]

Debye and his coauthors thus drew a suggestive picture of Born's qualification as a "full" physicist, who, like Voigt, would engage in experimentation and in the investigation of experimental data; they nevertheless stressed that it would not be possible for Born to replace Debye since "this was grounded in the rare double qualification of the outstanding scholar." It was in particular "a combination, which becomes untenable with the exit of professor Debye."[42]

Interestingly enough, nobody discussed the option of just hiring a theoretician, given the fact that Pohl had established splendid teaching of experimental physics. This is surprising, for Debye himself entered the Göttingen faculty through precisely this access opened up by a new and confident field of theory. We have seen how Hilbert never had tired of promoting this field of "theoretical research" for which the guest professorship was established. Now, only six years later and with the considerable success of theory, no traces of similar efforts can be found on Hilbert's part. The wind had shifted, mostly due to war experiences and postwar needs, but, clearly, changes in the faculty had also occurred. While Born had learned his lessons of applied science, theory testing and experimental insight during his war work, and in this way shaken off his former distance to experiential research, and had even started supervising work by students in this field, in Göttingen proponents of applied mathematics and their supporters from industry became more outspoken. As a consequence, Born would quickly find himself appeasing Felix Klein, to whom he always had a strained relationship, rather than mobilizing Hilbert as a local supporter.

After Born received the Göttingen offer in May 1920, he went to see Pohl first and had "intensive discussions" in order to define a "plan for the arrangements for the teaching of physics" that he then proposed to the ministry.[43] In a nine-page letter he took great pains to include all arguments and to devise a full architecture of posts and duties. Recognizing Pohl's great experimental lectures, which he performed with his personal apparatus, supporting his demand of promotion to an ordinary professorship and acknowledging his priority in using the large lecture theater, he, nonetheless, made clear that "the theoretician [also] has the right to lecture one or

---

[41]Document 18 (see Chap. 6).

[42]Draft Debye, Runge and Wallach, before 22 March 1920, faculty minutes, Phil. Fak. II Ph. Nr. 36 d, XV. The original reads: "Der jüngst hier verstorbene Prof. Voigt war ein Vertreter der theoretischen Physik von größtem Ruf. Wenn der als sein Nachfolger [...] bekannte Prof. Debye nach dem Tode des etatmäßigen ordentlichen Professors Riecke unter gleichzeitiger Anstellung des außerordentlichen Professors Pohl, in jenes Ordinariat einrücken und einen Teil der Aufgaben auch des Experimentalphysikers übernehmen konnte, so war das durch die seltene doppelte Qualifikation des hervorragenden Gelehrten bedingt und eine Kombination, die nun mit dem Ausscheiden des Professors Debye unhaltbar wird."

[43]Born to Dean (Hans Stille), 4 July 1920.

two hours every week in this theater, for there are occasional lectures that include demonstrations for which the small lecture room is not sufficient."[44]

Concerning the second position that Born wanted to give to James Frank, he had to walk a fine line between the claims of Pohl, as the master experimentalist, and the *Göttinger Vereinigung* championing technical physics. On the one hand, Franck would bring in his own approach to modern experimenting and with this was largely able to bridge theory and experiment; also, he had shared much of his career with Pohl and they had become good friends despite many differences in character.[45] Therefore, Pohl welcomed Born's suggestion in great unison with Hilbert, Runge and Prandtl. On the other hand, however, Klein raised objections: "he fears that technical physics would come off badly if a pure physicist is called for Simon's position," Born confided to Erich Wende of the ministry on Klein's opposition, "and the Göttingen Association for the Advancement of Applied Mathematical Physics could get one into trouble." Instead of drawing an artificial line between technical and pure physics, a distinction that is of no use when attempting to provide a broad education for their students, Born goes on, one should recognize: "There is only good and bad physics." And when it comes to finances, their support is "almost ridiculously small," so that he himself was quite confident to "to raise, if necessary, larger sums from industry off my own bat."[46]

Partly to appease Klein and the applied mathematics camp, partly to allow more personnel to be included, Born developed a detailed architecture both to span all fields of applied and experimental physics and to explain his elaborate space requirements. It involved Max Reich's interim position in applied electricity (though it hardly rendered him qualified to fill this position appropriately); Otto Stern to follow him suit from Frankfurt, if he did not become his successor there, in this way extending their productive collaboration; and Heinrich Rausch von Traubenberg, a lecturer who had shown a good command of teaching in some major experimental courses in 1917 and 1918. Born stated that "In the last years he gave the main experimental physics lectures—in part alone, in part parallel to with Pohl—in fact, with considerable success." and he concluded that, in this way, a "very promising order of instruction would result," which he specified meticulously with respect to teaching fields, duration, audience and suggested personnel (see Table 4.3).[47]

In the course of this differentiation of various kinds of experimental and applied physics, Born again began to downplay his own experimental aptitude. "I have never given experimental instruction, and I was able to get into this only when I abstained completely from my research work for several years," Born had explained to the

[44]Born to Wende, 15 Mai 1920, SBB Born papers, folder 1826, p. 2–14, on 2. "Ich gebe gern zu, daß der Experimentalphysiker Vorrecht auf den Hörsaal hat, weil er zum Aufbau seiner Versuche volle Freiheit haben muß. Aber ich halte es für gut, daß der Theoretiker das Recht hat, wöchentlich ein oder zwei Stunden in dem Hörsaal zu lesen; es gibt gelegentlich Vorlesungen mit Demonstrationen, für die der kleine Hörsaal nicht ausreicht."

[45]Cf. for the long relation between Franck and Pohl e.g. Ebner (2013).

[46]Born to Wende, 15 Mai 1920, SBB Born papers, folder 1826, p. 2–14, on 3.

[47]Ibid. p. 6.

ministry in his first long letter.[48] This was, however, only the first sketch for an entirely new program of physics teaching in Göttingen and marked the beginning of half a year of negotiations between Born, the two universities and the ministry.

Pointing to the Göttingen offer, which he reported to the head curator of the University of Frankfurt (who happened to be the mayor of Frankfurt, Georg Voigt, the one who had opened the university in 1914), Born listed his demands. They appear modest compared to those for Göttingen—Born had imparted earlier that he would "only reluctantly leave this fine city"[49]—but they still required a new full professorship for Stern, which was the only means to retain him as collaborator, a salary matching the Göttingen offer, and guarantees for subordinate personnel and funds for reorganization that were independent of "the fate of the University of Frankfurt," meaning provided by private funds or philanthropy.[50]

The young university, which had set up Laue's and thus Born's chair through the philanthropy of the Jewish jewelry trader and amateur astronomer Moritz Oppenheim, tried to meet all of Born's requirements but one, a position for Stern.[51] While no clear reason for this refusal can be found in the official correspondence and documents, Born gave Einstein an unequivocal explanation. Richard Wachsmuth, chair of experimental physics, rejected Stern because of his "corruptive, Jewish intellect," an attitude Born judged to be "at least open anti-Semitism," thus leaving much room for a variety of more veiled forms.[52] So, even in a rather liberal city like Frankfurt, Jewish participation hit its limits, which discriminated rather idiosyncratically between Born and Stern and their Jewishness. Things were not much different in Göttingen either, with Born and Franck entertaining a "ridiculous luxury of a private religion," so Pohl's view on Franck as conveyed to his mother in 1916 (Ebner 2013, 172). Born had to realize that he could neither win the uphill battle in Göttingen. Eventually, Stern became a full professor in Hamburg, an even younger university than Frankfurt, and, strangely enough, Frankfurt made some efforts to call Einstein to the Oppenheim chair, when open hostility against the creator of general relativity was staged in Berlin and had acquired anti-Semitic undertones.

Although Born told Voigt in Frankfurt back in early July 1920 that he was unable to turn down the Göttingen offer, it still took until November for him to sign the official agreement with the ministry on his Göttingen position and resources.[53] In this process some elements of the Göttingen architecture of teaching and personnel also shifted.

---

[48]Born to Wende, 15 Mai 1920, SBB Born papers, folder 1826, p. 2–14, on 5f.

[49]Born to Vorsitzenden des Kuratioriums der Uni Frankfurt, Oberbürgermeister Voigt, 18 May 1920, UA-Fra Personal-Hauptakte Born, Bl. 4.

[50]Born to Voigt, 7 June 1920, UA-Fra Personal-Hauptakte Born, Bl. 5.

[51]Voigt to Born, 3 July 1920, UA-Fra Personal-Hauptakte Born, Bl. 12f.

[52]Max Born to Einstein, 16 July 1920, and Hedwig Born to Einstein, 31 July 1920, (Born and Einstein 1969, 54–55).

[53]Born to Voigt, 10 July 1920, UA-Fra Personal-Hauptakte Born, Bl. 14. Agreement Born and Wende, 10 November 1920, GSPtKB Rep. 76 V a, Sekt. 6, Tit. IV, Nr. 1, Bd. XXVII, Bl. 76.

Table 4.3 A *Pflanzstätte* for mathematical-physical fields. Born's comprehensive 1920 plan for physics teaching in Göttingen, quotes arranged in table format (letter Max Born to Erich Wende, 15 Mai 1920, SBB Born papers, folder 1826, p. 6f.)

1.	Große Vorlesung Experimentalphysik (2 Sem., 4 St.)	Für alle Hörer	Ordinarius für Experimentalphysik (Pohl)
2.	Höhere Experimental- physik (2 Sem., 2 St.)	Solche Kapitel der Experimentalphysik, die sich für den allgemeinen Hörerkreis der großen Experimentalvorlesung nicht eignen, aber für Physiker und Mathematiker notwendig sind	Ordinarius für angewandte Elektrizitätslehre (Franck)
3.	Hauptvorlesung für theoretische Physik mit Übungen (5 St.)		Ordinarius für theoretische Physik (Born)
4.	Vorlesungen aus der angewandten Elektrizitätslehre (2 St.)		Ordinarius für angewandte Elektrizitätslehre (Franck)
5.	Elektrotechnik (2–3 St.)		Extraordinarius für Elektrotechnik (Reich)
6.	Spezialvorlesungen über theoretische oder experimentelle Physik (3–4 St.)		Extraordinarius für Physik (Stern bzw. Traubenberg)
7.	Radioaktivität und dergl. (2 St.)		Lehrauftrag für Radioaktivität (Traubenberg)
8.	Praktika (jedes Semester 4 St.)		
	a. für Nichtmathematiker (Mediziner, Pharmako- logen etc.)		Ordinarius für Experimentalphysik (Pohl)
	b. für Physiker und Mathematiker	- Elektrizität - Optik -Wärme - Atomistik, Elektronik	Ordinarius für Experimentalphysik (Pohl) Extraordinariat bzw. Lehrauftrag (Stern bzw. Traubenberg) Ordinarius für angewandte Elektrizitätslehre (Franck)
	c. für technische Physiker		Extraordinarius für Elektrotechnik (Reich)

Without consulting Born, the Göttingen dean pushed to fill the vacant professor- ship in applied electricity. As neither Reinhold Rüdenberg nor Heinrich Barkhausen was willing to come, only Max Reich was left; and this position was essential to maintain the support of the *Göttinger Vereinigung* for the university. Excusing his scant publication record with wartime secrecy that prohibited him from publishing his "most important and best work" as well as with his "grand modesty," however, the dean's request to appoint Reich was received with "slight surprise" by the ministry, for he had been added to the list only "with marked distance" to the aforementioned, and also because Born had not been involved. As officer Wende wrote to Dean Hans Stille, Born had already made different suggestions for the successor of Simon's position.[54] Stille, however, was able to reply that in the meantime Born himself considered this question "irrelevant" so that Reich should be appointed.[55]

Born's view on this matter was more nuanced, however, and it was rooted in his relationship to experiments. As he had always stressed that he could not teach experimental physics the way Debye did—though he might be willing to supervise doctoral students on experimental research as he had done in Frankfurt—the idea was to fill Simon's position not with a "technician" but rather with a physicist able to teach this field to students of mathematics and physics from a higher perspective. Here Franck would have been an ideal candidate. It was the *Göttinger Vereinigung* and the faculty that foiled this plan and, since he did not want to turn down the Göttingen offer, he had to negotiate again with the ministry. Consequently, in Berlin an agreement was reached that made it possible to hire Franck for an additional professorship anyway: Woldemar Voigt's former chair, formally degraded to an extraordinary professorship, was to be revived to a full professorship for Franck. In this way Born's plan could be realized if the necessary funds for experimental teaching would be provided.

Putting his own interests in heading an independent institute for theoretical physics last, Born suggested that he would be willing to make a double sacrifice: Under the condition that Franck receive a second full professorship for experimental physics, Born would both forgo the directorate in favor of Franck he would take on the full teaching load of theoretical physics alone. "But in this way," he wrote, "I acquire a collaborator with whom I share many common fundamental views and I expect great advantages for teaching and research from this."[56]

As a consequence, the division of resources with Pohl had now to be renegotiated with Franck, as Pohl was claiming the main lecture theater exclusively,[57] while Born tried to sell the new arrangement to Felix Klein, the principal proponent of the applied mathematics and technology fraction, which had at least successfully lobbied for Reich's appointment. Thanking him for supporting the appointment of Franck, Born wrote to Klein that he was fully aware of his responsibility for insisting on this very person, but he was quite sure that "Franck would, in fact, be able to create a school of grand style." Like Faraday, Franck had the rare gift to recognize experimental opportunities, a fact that had been appreciated abroad before it was "discovered" at home as Born had realized when he visited America in 1912. Franck was well aware of the Göttinger Vereinigung and would support it in the same

[54]Stille to Wende, 25 June 1920, UA-Fra Phil. Fak. II Ph. Nr. 36 d, XV. Wende to Stille, 2 July 1920, UAG, Phil. Fak. II Ph. Nr. 36 e, I.

[55]Stille to Wende, 7 July 1920, GSPtKB. Rep. 76 V a, Sekt. 6, Tit. IV, Nr. 1, Bd. XXVI, Bl. 418–419

[56]Born to Stille, 4 July 1920, UAG, Phil. Fak. II Ph. Nr. 36 e, I.

[57]Minutes "Nachfolge Simon" entry of session on 5 July 1920. = UAG Phil. Fak. II Ph. Nr. 36 e, I.

way as Born did in order to fulfill Klein's vision of science and industry stimulating each other. In addition, Born even told Klein of own plans for a "calculating institute for theoretical and technical physics," thus sketching common ground in some detail; it was a nicely crafted plan to Klein's liking, however, it was never realized in any form.[58]

In October and November 1920, agreements with the ministry redefined Göttingen physics into two experimental departments for Pohl and Franck, with two assistants each, and a mathematical department for Born, with one special assistant only. Nevertheless, Born secured 20,000 Marks—a considerable amount for apparatus, thus highlighting his continuing interest in experimental research.[59] This appears less surprising given the fact that at this time Born was in the middle of his own experimental work in Frankfurt and the Stern-Gerlach experiment was at a crucial stage. Born and his Frankfurt collaborators had just presented their experimental work prominently at the *Naturforschertagung*, which took place in Bad Nauheim in September 1920 (historically remembered mostly for the dispute between Philipp Lenard and Albert Einstein on relativity) (Born 1920b; Stern 1920).

Later that year Born wrote Klein about his preparations to move to Göttingen for the spring semester: "I am very busy as it means a lot to me to finish an experimental project before my departure."[60] Against this background of experimental work in Frankfurt, where he had a number of students performing experimental work but none on theory, it appears consistent that Gerlach was granted the status of *Privatdozent* without having fulfilled any of the standard requirements, and that he was giving his inaugural lecture on "The importance of quantum theory in modern experimental physics."[61]

In late 1920 and early 1921, a number of shifts in disciplinary terrain and developments in physics accompanied the process of Born's and Franck's relocation to Göttingen. In arguing to raise the pay for his "physics assistant," who had received an offer for an assistant position from Frankfurt (probably from Madelung, who was about to be named Born's successor there, while Elisabeth Bormann had requested to be released from her contract),[62] Hilbert claimed that it was this assistant, Dr. Kratzer, "on whose shoulders presently rested essentially all activity of theoretical physics here." Without better funding of his assistant positions, Hilbert feared, his teaching would be at risk, which covered wide areas of physics as has been detailed above, a field Born would quickly reclaim, however.[63] Before, a lectureship in the

---

[58]Born to Klein, 11 July 1920, SUB Göttingen, Klein Papers, 5 C Bl. 68/69.

[59]Agreement Wende and Franck, 6 October 1920, GStA PK. Rep. 76 V a, Sekt. 6, Tit. IV, Nr. 1, Bd. XXVII, Bl. 63–64. Agreement Wende and Born, 10 November 1920, GstA PK. Rep. 76 V a, Sekt. 6, Tit. IV, Nr. 1, Bd. XXVII, Bl. 76.

[60]Born to Klein, 21 November 1920, SUB Göttingen, Klein Papers, 5 D, Bl. 79.

[61]Dean Lorenz to Kurator, 8 February 1921, UA-Fra Personal-Hauptakte Gerlach, Bl. 9.

[62]Bormann to Curator, 2 April 1921, UA-Fra Personal-Hauptakte Born, Bl. 6.

[63]Hilbert to Minstry, 10 November 1920, GstA PK. I. HA Rep. 76, Nr. 591, Bl. 280.

field of (experimental) radioactivity had been granted to Rausch von Traubenberg, in full accordance with Born and Franck, as the Göttingen dean wrote to the ministry, thus extending a line of practical classes that Eduard Riecke had established long ago.[64]

4.3 Quantum Theory in the Göttingen Curriculum, 1921–1926

A striking difference to academic conditions in Frankfurt was apparent for Born after the first days of teaching in Göttingen. He reported to Gerlach that together with Hilbert, Franck and Runge he had to supervise 80 students, including some quite capable ones, and in the seminar that he offered together with Franck there were even more. He invited Gerlach to join him in Göttingen "for the purpose of leisure: Monday seminar, Tuesday mathematical seminar, Wednesday mornings proseminar, Wednesday afternoon colloquium."[65] Therefore he opted to teach, on the one hand, about topics in which he had firmly established himself, in this case four-hour courses on kinetic theory as he was just completing the second edition of his book on this topic with the help of a private assistant (Emmerich Brody); and, on the other hand, standard fare any physics student had to master: mechanics, optics, heat etc. (see Table 4.4). Clearly, Born's grand plan for teaching physics, developed and negotiated with Pohl and Franck, now had to be met, which restricted Born's latitude for teaching. In consequence, those students interested in recent quantum and atomic physics in 1921 had to approach this field from the experimental point of view through Franck's two-semester lecture course on atomic physics, since no lectures on atomic theory were offered. Hilbert and lecturer Paul Hertz at least were offering other fields of modern physics, relativity and radiation theory, Hilbert's course, however, was advertised as "basic ideas of relativity theory (for students of all fields)" and was thus a popular treatment just one hour per week, rather than an advanced course.[66] Solid physical theory was then offered by Hilbert in the following term with a weekly four-hour lecture on "statistical methods, particularly in physics" which also included some quantum theory of specific heats (Hilbert 1911, 104).[67]

Coincidentally or not, after summer 1922 when Niels Bohr delivered his famous talks on atomic physics, a total of seven spanning two weeks,[68] both Hilbert

---

[64]Stille to Ministry, 23 August 1920, GstA PK. Rep. 76 V a, Sekt. 6, Tit. IV, Nr. 1, Bd. XXVII, Bl. 7.

[65]Born to Gerlach, 11 May 1925 and 16 May 1925, DMA, Gerlach papers 83. "Kommen Sie mal auf ein paar Tage zur Erholung nach Göttingen."

[66]"Grundgedanken der Relativitätstheorie (für Hörer aller Fakultäten)," the lecture notes went under the title "Über Geometrie und Physik" (by Bernays) and "Über Geometrie und Physik" (by Hückel).

[67]Statistische Mechanik (listed as "Statistische Methoden, insbes. der Physik"), summer term 1922, lecture notes by Lothar Nordheim, Mathematisches Institut Göttingen. [SS 1922], Mathematisches Institut Göttingen.

[68]Lecture notes of 12, 13, 14, 19, 20, 21 and 23 June 1922, SBB, Born papers 1819.

Table 4.4 Lectures and joint seminars of Hilbert, Born, Franck and others 1921–1926. Selection with respect to quantum theory and new topics in physics, philosophy and methodology. According Verzeichnis der Vorlesungen auf der Georg-Augusts Universität zu Göttingen

/	Hilbert	Born	Franck	Other
SS 21	Einsteinsche Gravitationstheorie Relativitätstheorie*	Kin. Theorie der festen Körper I: Elektrizität, Magnetismus	Atomphysik I	*Hertz* Strahlungs-theorie
		Struktur der Materie ǀ Physikalisches Proseminar		
WS 21/22	(Grundlegung der Mathematik)	Kin. Theorie der festen Körper II: Elektronen-theorie, Optik	Atomphysik II	*Hertz* Statistische Mechanik
		Struktur der Materie ǀ Physikalisches Proseminar		
SS 22	Statistische Metho-den insb. der Physik	Theorie der Wärme Elektro- und Magnetooptik	Elektronenleitung durch Gase	
		Struktur der Materie ǀ Physikalisches Proseminar		
WS 22/23	Math. Grundlagen der Quantentheorie Wissen und mathema-tisches Denken*	Kin. Theorie d. Materie	Radioaktivität	*Courant/Siegel* Differenzengl.
		Struktur der Materie ǀ Physikalisches Proseminar		
SS 23	(Anschauliche Geometrie)	Mechanik	Atomtheorie	*Hertz* Methodenlehre
		Struktur der Materie ǀ Physikalisches Proseminar		
WS 23/24	(Mengenlehre)	Höh. Mechanik[a]		
		Struktur der Materie ǀ Physikalisches Proseminar		
SS 24	Mechanik	Atommechanik II	Grenzgebiete d. Phys. u. Chemie	*Hertz* Prinzipien...
		Struktur der Materie ǀ Physikalisches Proseminar		
WS 24/25	Das Unendliche*	Optik	Elektriziätsleitung durch Gase	
		Struktur der Materie ǀ Physikalisches Proseminar		
SS 25	(Ans. Geometrie)	Theorie der Wärme	Exp. Grundl. d. Quantentheorie	*Hertz* Methodenlehre
		Struktur der Materie ǀ Physikalisches Proseminar		
WS 25/26	(Zahlentheorie) Raum und Zeit*	Kin. Th. der Materie	Exp. Grundl. d. kin. Theorie	*Heisenberg* Kristallgitter
		Struktur der Materie ǀ Physikalisches Proseminar		
SS 26	(Algebr. Zahlen)	Atomtheorie	Fluoreszenz u. Phosphoreszenz	*Hertz* Methodenlehre
		Struktur der Materie ǀ Physikalisches Proseminar		
WS 26/27	Math. Methoden der Quantenth.	Elektrizität und Magnetismus	—	*Hertz* Methodenlehre
		Struktur der Materie ǀ Physikalisches Proseminar		

SS summer term, WS winter term

*Lecture open to students of all fields

( ) Lectures on other topics mentioned only if no other relevant lecture was offered

[a] Added in brackets on lecture notes: "Störungstheorie mit Anwendungen auf Atomphysik."

and Born offered courses covering quantum topics. Again, Born was approaching radiation and quantum theory from the point of view of his kinetic theories, thus excluding Bohr's theory and atomic modeling, while Hilbert, in his typical way, used his lecturing to explore new research fields and secured help from two physics assistants, Lothar Nordheim and Gustav Heckmann, both advanced doctoral students of Born (Heckmann 1924).[69] These lectures presented the Bohr-Sommerfeld theory in a mathematically somewhat streamlined version, including Bohr's quantization rules, the framework of Hamilton-Jacobi theory and action-angle variables, and it closed with research results as recent as Born's and Pauli's paper of 1922 (Born and Pauli 1922; Hilbert 2009a, 503–602). In the following term, only Franck offered a course titled "atomic theory," although his teaching was otherwise always on experimental physics.[70]

Only from winter term 1923/24 on did Born adopt Hilbert's teaching philosophy and used his courses on "higher mechanics" and "atomic mechanics II" to systematically treat the application of astronomical perturbation theory on atomic physics. With the assistance of Hund, again Nordheim and also Heisenberg for some parts the lectures were published in early 1925 as "lectures on atomic mechanics," and were intended, as indicated in the preface, as an advanced treatment that would significantly differ in particular from Sommerfeld's book, though the student should find this out himself (Born 1925; Sommerfeld 1922). At this time Born also had monopolized modern theoretical physics, as Hilbert would not return to teach relativity until 1925 and quantum theory until 1926, concentrating instead on central topics from mathematics, and Paul Hertz had also shifted his field to the methodology of natural science.

Regarding physics, one finds a curricular pattern and a division of labor between Born and Franck which was repeated in an approximately two-year rhythm, while Hilbert, too, duplicated and extended his 1922/23 course on the mathematical foundations of quantum theory four years later, now including the fully developed quantum mechanics (Hilbert 2009a, 605–707).

The strongest connection between teaching and research was made in the seminars on which, however, no contemporary sources exist aside from some memoirs. What seems obvious from Table 4.4 is that Born had established himself as the central actor, who participated actively both in the famous structure of matter seminar with Hilbert and in Franck's seminar, which was meant as a forum where students of experimental and theoretical physics could inform each other about their research work on their dissertations.

---

[69]Nordheim's dissertation was published in Nordheim (1923a) and Nordheim (1923b).

[70]No lecture notes on this course are known to have survived.

Chapter 5
Göttingen's Multiple Avenues Towards Quantum Mechanics

For Born's and thus for Göttingen's route to quantum mechanics personal experience in experimental physics was an essential ingredient as was the close monitoring of experimental results relating to quantum effects in atomic physics like the Ramsauer effect, which was widely discussed in Göttingen (Ramsauer 1921). It meant a return to more mathematical work when Born embarked on for the application of celestial mechanics to the atom from 1922 on, first with Wolfgang Pauli and then with Werner Heisenberg.

However, comparison of the results with experimental measurements remained key. In a letter inviting to join the Bohr talks in May 1922, which were paid by the Wolfskehl fund, Born entered the contest with the Munich school: "And otherwise, I let my students quantize, just to compete a little bit with you."[1] Here, on the one hand, he would soon outdo his Munich colleague in mathematical sophistication—using the rich theories Hilbert and his associates had prepared for him; on the other hand, it was Born who led his younger Göttingen collaborators back to experimental realities, by introducing the probability interpretation and relating the abstract quantum theory to scattering experiments.

In the end, it proved successful that he had kept rather close contact with experiments, very much so with his friend and colleague James Franck, to whom he entrusted the experimental part of the Göttingen position which he had initially been expected to take on, too. In this way, Born was able to invest most of his time and effort in theory, but he maintained close attention to experimental findings and concepts, among them, the concept of collisions in particular. As late as 1924 Born supervised a student on molecular beam experiments employing and refining equipment, which he brought from Frankfurt and immediately had reinstalled in Göttingen.[2]

---

[1]Born to Sommerfeld, 13 May 1922; (Sommerfeld 2004, 118f.) "Auch sonst lasse ich meine Leute quanteln, um Ihnen ein wenig Konkurrenz zu machen."

[2]Bielz (1925) summarizes the Göttingen dissertation of 1924.

© The Author(s), under exclusive license to Springer Nature Switzerland AG 2019
A. Schirrmacher, *Establishing Quantum Physics in Göttingen*,
SpringerBriefs in History of Science and Technology,
https://doi.org/10.1007/978-3-030-22727-2_5

5.1　Born's and Franck's Research Agendas

When Alfred Landé received news from Bohr in Copenhagen early 1921, that he was now able to derive the whole periodic table from his quantum theory, Born (who was still in Frankfurt), Gerlach and Stern, wondering how this could be possible, urged Franck to worm the secret out of Bohr whom he was about to visit.[3] Plans to make Bohr come to Göttingen for Wolfskehl lectures in spring were not realized until the following year, when, at the "Bohr festival" Heisenberg also entered the stage and began working with the new Göttingen quantum physics community spearheaded by Born and Franck. This means that Born and Franck were setting up their research agendas and working groups for at least a year before Bohr was to deliver his most influential yet somewhat puzzling and vague talks.

Instead of interpreting this visit as a "dramatic event," that started a "new Göttingen era,"[4] in the following, I will instead stress a number of continuities, which are related to Born's collaborations, specific experimental equipment or findings, and methods used. At least five such lines of research fields and agendas can be identified: X-ray structural analysis, molecular beams, many-electron quantum theory, the Ramsauer effect and collision theory. My claim is that these five strands of physical science, that were experimental, theoretical or often rather a combination of both, constitute the Göttingen avenues to quantum mechanics. Interestingly, they did not meet exactly in one point—matrix mechanics—but displayed various crossings of different nature.

Exactly this finding also brings us back to the perspective of resources and research politics and the historical observation that the actual trajectory of scientific progress is the result of many decisions to direct or redirect resources in a way that favors or encumbers research in a particular field. As we have seen, this often entails resources of various kinds like constraints of finances or personnel, the investment of time of the single scientist or the promise of the given research project within science or as an exploitable technology.

Relocation of Research and Acquisition of New Resources Including X-rays

Born's move to Göttingen did not mean an end to the experimental research he had initiated in Frankfurt; on the contrary, a remarkable continuity can be found. When he wrote to Gerlach in May 1921 that he liked Göttingen and that his and Franck's assistants were busy assembling "my apparatus" quickly, with only liquid air still missing, at the same time he was arranging for the work of his old collaborators to proceed. In the case of his student Peter Lertes, who had already completed his dissertation at the university in 1920 but was now improving his experimental work for publication in *Zeitschrift für Physik*, Born made clear that he could easily duplicate

---

[3]Born, Gerlach and Stern to Franck, 22 February 1921, SBB Born papers, folder 954, pp. 1–2.

[4]For the interpretation as a "dramatic event" see e.g. Mehra and Rechenberg (1982, Chap. III, 262). That Bohr's talks did not present much spectacular news can be be inferred from a comparison of the notes taken of the event (SBB Born papers, folder 1819) and the published Copenhagen lectures, which had been submitted to Zeitschrift für Physik back in January 1922 (Bohr 1922).

his work on dielectric fluids in Göttingen (Lertes 1920, 1921a, b): "Please tell Lertes," he wrote,

> when he annoys you with his laziness etc. I will have his rotation experiment built up in Göttingen; here are many people who can handle amplifier tubes and could do the job quickly. One has to talk to the stripling à la Entente in short-term ultimatums with sanctions.[5]

More cordially, Born praised Gerlach's energy to push forward crystal analysis, a project for which Born had provided the theoretical computations. Now he was working with him to figure out how well the atomic theory could be confirmed and where things still had to be improved. Commenting a recent paper by Born's new Göttingen collaborator Franck, Born and Gerlach compared various experimental and theoretical accounts for electron affinities concluding that the force law, which describes the atomic repulsion within a lattice, needed refinement (Born and Gerlach 1921a). A week later, Born besought Gerlach, "Beware of X-ray burn!!!!!!" and threatened to write his wife because any injuries would be irreversible and what he had heard from his assistant about Gerlach's carelessness really scared him.[6]

But not only did projects in Frankfurt extend to Göttingen, in a certain manner, Debye's research here before Born's arrival had spread to Frankfurt as well. It was precisely these measurements with X-rays and the Deybe-Scherrer camera setup what made Born worry about Gerlach's lab practice and safety. Some of the equipment for this work was provided by the Solvay Institute (with von Laue's help an X-ray transformer was borrowed) and some by private funding that Born had secured (Gerlach and Pauli 1921). At the end of 1921, Gerlach was involved in at least three lines of experimental research: besides Debye-Scherrer type X-ray diffraction, he also pursued molecular beams, which became the Stern-Gerlach experiment, and optical activity, all fields that offered stronger or weaker cooperation with Born. In October Born even suggested investigating a new effect when Gerlach had started looking into an experimental method to determine the impact of atomic dipoles on (mechanical) friction coefficients. He conjectured that rotating dipoles should produce measurable magnetization.[7]

Born was aiming at much more upscale X-ray equipment for Göttingen, and his closeness to Einstein was instrumental in gaining access to funding from the Kaiser Wilhelm Institute for Physics, which was still a rather virtual institution. As the respective details have been discussed in Castagnetti and Goenner (2004), it suffices here to mention that Born, on the one hand, was strengthening his case by siding with Franck and Pohl—eventually, they made the request jointly. On the other hand, he was able to forgo expounding definite research needs in the application, in this way winning the allocation of funds of no less than 100,000 Marks (although inflation had started to accelerate).[8] Thanking Einstein for the great gift, he explained that such

---

[5]Born to Gerlach, 16 May 1921, DMA Gerlach Papers 83.

[6]Born to Gerlach, 23 May 1921, DMA Gerlach Papers 83.

[7]Born to Gerlach, 10 Oct 1921, DMA Gerlach Papers 83.

[8]Verzeichnis der im Rechnungsjahre 1921/1922 bewilligten Zuwendungen, Archiv der Max-Planck-Gesellschaft Berlin, I. Abt. Rep. 34, Nr. 13.

equipment "is now part of a reputable institute and there are quite often questions that can only be answered with the help of X-rays. [...] This grand donation means that all of you in Berlin are confident that we might figure out something respectable with the apparatus, which makes us happy."[9]

Molecular Beams

One particular line of experimental research that extended from 1920 up to 1925 and came from Frankfurt to Göttingen (and with Otto Stern to Hamburg) was atomic or molecular beams. When Born referred to "my apparatus" to be reassembled in Göttingen, he meant the experimental setup he had used for the determination of the mean free length of silver atoms with some help from Elisabeth Bormann (Born 1920b; Born and Bormann 1921). Although not much more is known about the use of this equipment in the first years in Göttingen, a doctoral student of Born, Fritz Bielz, explained at some length in *Physikalische Zeitschrift* in 1925 how the "experimental setup had been improved by Prof. Born in cooperation with Dr. E. Hauser and Dr. R. Minkowski," two assistants from Göttingen and Hamburg, respectively. In this way, he was able to realize useful quantitative measurements including the measurement of the diameter of an uncharged silver atom. Thanking Born both for suggesting and for supporting the execution of the project, and also the two mechanics of the institute for constructing the somewhat complicated brass apparatus, Bielz' dissertation is a good example for Born's continued interest in experimental atomic physics (see Bielz 1925 and Table 5.1).

Many-Electron Quantum Theory

Contrary to standard historiography, Born's program to systematically push forward Bohr's quantum theory for systems with few or many electrons was not a result of the 1922 "Bohr festival," nor did it coincide with his remark on competition towards Sommerfeld. Rather, it can be traced back to Born's late Frankfurt days. Born wrote Einstein in February 1921 that he had just hired a private assistant—like Hilbert used to acquire his physics assistants—to help him complete his book on the atomic theory of the solid state (Born 1923).[10]

Emmerich Brody was a Hungarian Jew with "Eastern manners" and weak hearing and thus appeared in Born's judgment untenable for the German university system, so he was paid from private funds.[11] It was within this collaboration that in the context of a theory of crystal lattices both the problem of anharmonic oscillations and the use of perturbation theory, adapted from celestial mechanics, were brought to bear on quantum theory. Both were crucial resources on which Born later drew in work with

---

[9]Born to Einstein, 29 November 1921, Born and Einstein (1961, p. 91ff).

[10]A critical discussion of the applicability and the limits of Bohr's theory can be found already in this book, including the untenability of flat orbital models, the unobservability of electron rings by X-ray analysis and that atomic dynamics may relate to infinite-dimentsional quadratic forms (matrices) which may have both a discrete and an continuous spectrum, Cp. pp. 712, 754, 594f.

[11]Born to Einstein, 12 February 1921, Born and Einstein (1969, 81, 100). On Brody, see Born (1975, 214).

Table 5.1 Dissertations concluded under Born's supervision in Göttingen 1921–1925

Name	Topic	Year	Published in
Hund, Friedrich	Versuch einer Deutung der großen Durchlässigkeit einiger Edelgase für sehr langsame Elektronen	1923	Z. Phys.
Hermann, Carl	Über die natürliche optische Aktivität von einigen regulären Kristallen (NaClO$_3$ und NaBrO$_3$): eine Prüfung der Bornschen Theorie der Kristalloptik	1923	Z. Phys.
Peter, Fritz	Über Brechungsindizes und Absorptionskonstanten des Diamanten zwischen λ 644 und 266	1923	Z. Phys.
Heckmann, Gustav	Über die Elastizitätskonstanten der Kristalle	1923	
Schmick, Hans	Zur Theorie der Dipolflüssigkeiten	1923	Z. Phys.
Nordheim, Lothar	Zur Behandlung entarteter Systeme in der Störungsrechnung/Zur Quantentheorie des Wasserstoffmoleküls	1924	Z. Phys.
Kornfeld, Heinz	Über die Bindung der Partikel in den Gittern verschiedener Dipol- und Quadrupolgase	1923	
Wessel, Walter	Über das Massenwirkungsgesetz in ionisierten Systemen und die numerische Berechnung chemischer Gleichgewichte	1924	
Jordan, Pascual	Zur Theorie der Quantenstrahlung	1925	Z. Phys.
Bielz, Fritz	Versuche zur direkten Messung der mittleren freien Weglänge' von ungeladenen Silberatomen in Stickstoff	1925	Z. Phys.
Bollnow, Otto	Zur Gittertheorie der Kristalle des Titanoxids, Rutil und Anatas	1925	Z Phys.
Bubenzer-Rolan, Karl	Die Eigenschwingungen tetraederförmiger Molekeln	1925	Z. Phys.

Pauli and Heisenberg, while Schrödinger found the application of quantum theory unnecessary here.[12]

Born, who had waived claims for one or more regular assistants in order to reach a deal with Franck and Pohl, had only one special assistant, Hans Weigt, who had done his dissertation with Debye on the electrical moments of molecules and whose skills were greater in the experimental fields (Weigt 1921). In July 1921—again, long before what is usually suggested to be the starting point of Born's program toward a quantum mechanics—Born urged the university to get Pauli hired as his assistant, since "Mister Pauli is regarded to be the greatest talent that has emerged in the field of physics in recent years" and the wish to collaborate with him had also been expressed by Hilbert, Runge, Franck and other colleagues.[13] His highly ingenious, though eventually unsatisfactory treatment of the simplest atom with two electrons, the hydrogen molecule ion, complemented efforts by Born and Hilbert (Pauli 1922).

---

[12]Born and Brody (1921), Born (1922b), Born and Pauli (1922), Born and Heisenberg (1923), Brody (1921), Schrödinger (1922).

[13]Born to Curator, 4 July 1921, UAG 4 V h 35, Bl. 185.

Both supervised Lothar Nordheim's dissertation on degenerate systems in quantum theory and the very same problem of the hydrogen molecule ion, treated in a different fashion (Nordheim 1923a, b).

This line of research included Brody, Pauli, and after his departure, also Heisenberg, who had come to Göttingen in fall 1922 still as a student (working on his dissertation about a problem of hydrodynamics) to learn from Born. After a first joint paper dealing with a model of Landé, Born and Heisenberg also dealt with the Helium problem, now the full, though excited atom, concluding that the principles of the old quantum theory developed by Bohr and Sommerfeld even when extended by methods adapted from celestial mechanics fail for deriving realistic properties (see Table 5.2).[14]

All these collaborations led to some preparatory efforts and results which Born understood as the attempt to push the mechanical picture, the quantization and the approximation methods to the very limit, in order to recognize those elements of the classical theory that ultimately fail to provide appropriate models consistent with the experimental properties of the real systems studied (Born 1924, 1925). Concerning Heisenberg, it may be noted that his funding followed very much the lines Voigt and Hilbert had pioneered: After receiving his doctorate in Munich, he was able to return to Göttingen in the autumn of 1924, funded in half by a grant from the Electrophysics Committee of the Notgemeinschaft, which Born had organized—in this way diverting money from industry that was meant to help technical physics towards theory—and in half paid directly by Born to make it a full assistant's salary.

Ramsauer Effect

As Gyeong Soon Im discussed in his pioneering study on the experimental constraints on formal quantum mechanics, another experimental input played a huge role in the discussions in Göttingen about solving the quantum riddle in the early 1920s: the Ramsauer effect (Im 1996). Originating from a rather different culture of physics, Philipp Lenard's Heidelberg institute with its long tradition on cathode ray research, the effect that slow electrons could, contrary to classical theory, penetrate noble gases almost freely, despite strong inter-atomic force fields, invited a response from quantum theory (Ramsauer 1921).[15]

Immediately after having heard Carl Ramsauer's talk at the Jena *Physikertag*, Franck wrote to Bohr that he was surprised by "a paper by Ramsauer that I am not able to believe, though I cannot show any mistake in the experiment. Ramsauer obtained the result that in argon the free path lengths are tremendously large at very low velocities of electrons. [...] If this result is right, it seems to me fundamental."[16] Born, in turn, told Einstein about his reaction to Ramsauer's presentation with respect to the theoretical consideration of the mean free length on electrons in gases: "[T]his, however, is of interest because of the simply crazy claim by Ramsauer (at Jena) that

---

[14]David Cassidy's description of Heisenberg's Göttingen time still gives a good overview about the collaborations and publications (Cassidy 1992).

[15]For a discussion from the experimental perspective, see Im (1995).

[16]Franck to Bohr, 25 September 1921 (Bohr CW, Vol. 8, p. 689).

Table 5.2 Born's collaboration in various fields of quantization, atomic structure, molecule formation and quantum collision theory, 1921–1926 (selection). Dates are those when received by the jounnal, or of publication date (marked with *), when not available

Authors	Title	Date
Born + Brody	Über die Schwingungen eines mechanischen Systems mit endlicher Amplitude und ihre Quantelung	09/07/21
Born + Pauli	Über die Quantelung gestörter mechanischer Systeme	29/05/22
Born	Über das Modell der Wasserstoffmolekül	27/06/22
Born + Brody	Zur Thermodynamik der Kristallgitter II	26/08/22
Born + Hückel	Zur Quantentheorie mehratomiger Molekülen	01/11/22
Born + Heisenberg	Über Phasenbeziehungen bei den Bohrschen Modellen von Atomen und Molekülen	16/01/23
Born + Heisenberg	Die Elektronenbahnen im angeregten Heliumatom	11/05/23
Born	Quantentheorie und Störungsrechnung	06/07/23*
Born + Heisenberg	Zur Quantentheorie der Molekeln	21/12/23
Born + Heisenberg	Über den Einfluß der Deformierbarkeit der Ionen auf optische und chemische Konstanten	10/02/24
Born + Franck	Bemerkungen über die Dissipation der Reaktionswärme	07/10/24
Born	Die chemische Bindung als dynamisches Problem	26/12/24*
Born + Franck	Quantentheorie der Molekelbildung	15/12/24
Born + Jordan	Zur Quantentheorie aperiodischer Vorgänge	11/06/25
Heisenberg	Über quantentheoretische *Umdeutung* kinematischer und mechanischer Beziehungen	29/07/25
Born + Jordan	Zur Quantenmechanik	27/09/25
Born + Jordan + Heisenberg	Zur Quantenmechanik II	16/11/25
Born + Jordan + Nordheim	Zur Theorie der Stoßanregung von Atomen und Molekülen	27/11/25*
Born + Wiener	Eine neue Formulierung der Quantengesetze für periodische und nichtperiodische Vorgänge	05/01/26
Nordheim	Zur Theorie der Anregung von Atomen durch Stöße	14/02/26
Born	Zur Quantenmechanik der Stoßvorgänge (Vorläufige Mitteilung)	25/06/26

in argon the free length becomes infinite with decreasing velocity (the atoms are penetrated by *slow* electrons!). This, however, we would like to disprove."[17] In this way, the Ramsauer effect contributed to bringing the problems of quantum theory to center stage in Göttingen.

Here this problem was tackled from both sides, experiment and theory, and Born and Franck were in close exchange. Franck's assistant Herta Sponer, a 1920 doctoral graduate of Debye in theoretical physics, had turned to experiments on atomic collisions after Franck had come to the university. When she published results from her

---

[17]Born to Einstein, 29 November 1921, Born and Einstein (1969, 92f).

experiments, first together with Rudolf Minkowski from Hamburg and then alone, Friedrich Hund had just finished his dissertation with Born on a theory of the Ramsauer effect.[18]

Hund had initially started from a purely classical approach and only turned to quantum theoretical arguments after Frank, during a stay at Copenhagen, came to the conclusion that the Ramsauer effect is even a requirement of quantum theory as soon as electrons have slow velocities. Frank's general idea that the correspondence principle might allow solving the problem could not be demonstrated by Hund, who was only able to present a qualitative derivation of the effect mostly dropping quantum theory and resorting to classical radiation theory with some additional constraints. As Born and Franck judged Hund's dissertation as proof that a classical explanation of the Ramsauer effect was not feasible, the work did not provide much of an answer.[19] Also Sponer and Minkowski concluded: "A possible interpretation of the effect could be given in relation with Hund's theory. Due to its uncertainty, however, we would rather not address this interpretation for the time being" (Minkowski and Sponer 1923).

It was precisely at this stage that Born started a new attempt to solve the Ramsauer puzzle. With the help of Pascual Jordan, who had succeeded Pauli as Born's assistant after his return to Hamburg, Hunds' early qualitative results should be replaced by a solid and detailed treatment based on a quantum theory of aperiodic processes. It is here that Bohr's correspondence principle, which Franck had already mobilized, was invoked more thoroughly and sharpened (Born and Jordan 1925b, 504). The extent to which the problem of the Ramsauer effect motivated research and the introduction of new concepts into the description of collision processes more generally may also illustrate the proposal by Walter Elsasser to perform diffraction experiments with free electrons according to de Broglie's dissertation. Initially a student of Franck, who then turned to theory and Born, Elsasser learned from both that such experiments had just been reported and that, by and large, the magnitudes would agree, so that he should now work things out theoretically (Jammer 1966, 249; Born and Einstein 1969, 121; Im 1996; Elsasser 1925, 711).

Collision Processes

For those like Born and Franck who were less interested in spectroscopy, for which the quantum theory of periodic processes was quite effective, but rather immersed in experiments of particle rays and molecular beams—a field that would eventually grow into the huge industry of particle physics—it was imperative to find a quantum theory that would not only deal with periodically orbiting electrons within the atom, but also with collisions. In their still pre-quantum mechanics paper, Born and Jordan started from the classical theory for aperiodic electrical fields acting on a restricted periodic system. They discussed induced probabilities of quantum jumps and employed a

---

[18] Hund (1923), Minkowski and Sponer (1923), Sponer (1923); cf. also the summary (Minkowski and Sponer 1924).

[19] Cf. Jähnert (2015), where Hund's and Franck's work on the Ramsauer effect is rather seen as "peripheral for the development of matrix mechanics or a successful quantum theory of collision processes" (p. 212); on Franck's collaborations and discussions see Lemmerich (2007, 116).

"correspondence principle of movements" in order to explain the Ramsauer effect with quantum theory. This meant the use of a technique, which can be traced back to Hilbert's lectures on quantum theory in winter 1922/23, and which was also discussed by various Göttingen physicist concurrently: the use of averaging methods and the replacement of differential by difference equations.[20]

As Born explained to Einstein on July 15, 1925, the motivation for all this work was to understand collision processes, which Franck had mastered in Göttingen:

> Chiefly, however, I am interested in the also quite mysterious difference calculus that is behind the quantum theory of atomic structure. With Jordan, I am systematically examining—though with limited mental effort—every imaginable correspondence relationship between classical, multiply periodic systems and quantum atoms. [...] This is preliminary work for an investigation into the processes at atomic collisions (quenching of fluorescence, sensitized fluorescence à la Franck, etc.); [...] (Born and Einstein 1969, 118f.)

Born and Franck were in ongoing discussions about collision processes from the very beginning of their Göttingen collaboration and in the crucial years of the formulation of matrix mechanics of Born, Heisenberg and Jordan, which, however, was not able to cover these processes before wave mechanics came to its aid. Born was alternately publishing with Jordan and Heisenberg on a new mechanics of matrices and with Franck on the formation of molecules from collision processes (Born and Jordan 1925b).[21]

With Franck he explored the question of molecule formation (and dissociation) and asked what role excitation and collision processes may play. They concluded that one could not understand molecule formation processes as collisions of just two atoms or ions, but rather, as energy considerations suggested, collisions of three involved entities. Born's theoretical analysis showed that quantization had to distinguish rather slow nuclear rotation and oscillation from the motion of the faster electrons, an idea Franck had already developed rather intuitively and qualitatively.[22] What later became known as the Franck-Condon principle, and was then fully derived from quantum mechanics (Condon 1926), described the possible molecule formations and processes in which energy differences of configurations would be balanced not by radiation energy but by translation, rotation or vibration, i.e. mechanical energy. These radiation-free processes where the released quantum energy changed into kinetic or translational energy of the atom without radiation were also called

---

[20] There is a priority debate in the secondary literature about this point, typically attributing invention of this technique to the respective person treated, i.e. Heisenberg by Jammer (1966), Mehra and Rechenberg (1982), Cassidy (1992), Kramers by Dresden (1987), or Van Vleck by Duncan and Janssen (2007). All their datings are preceded by the treatment in Hilbert's lectures, which have widely been heard and studied as the lecture notes were available (Hilbert 2009a, 503–602). This may be a typical example of the problems with dating a key idea which was, however, essentially developed collectively.

[21] Received 11 June 1925. For a detailed description of Franck's research and his collaboration with Born cf. Lemmerich (2007, 82ff.).

[22] Franck followed (Klein and Rosseland 1921; Franck 1922, 1924; Born and Franck 1925).

collisions of the second kind, and they were at the heart of many projects by Franck's doctoral students.[23]

5.2   Matrix and Quantum Mechanics in Göttingen

If the above considerations and the approach of considering the available and mobilizable resources in Göttingen as well as the presentation of the multiple avenues towards quantum mechanics were convincing to some extent, it should have become clear that they in fact develop a historical explanatory power, which is perhaps reinforcing, if not complementary to the prevalent conceptual approaches. Clearly, there is a wide literature reconstructing the detailed sequence of, and influences between scientific publications or the direct and written exchanges (though not so much about the laboratory procedures, both with experiments and theoretical ones like joint calculating, etc.),[24] and this literature is still evolving, for example, with meticulous reconstructions of the development of ideas like the correspondence principle, dispersion theory or Heisenberg's *Umdeutung*.[25]

My approach of analyzing the role of the institutional and resource regimes for the establishing of quantum theory in Göttingen is a first attempt and rather singular example for such a multi-layered contextual historiography. However, as the sketch of the collaboration between Born and Franck in the last section has demonstrated, there were many strong experimental constraints that Born was fully aware of when he was formulating the theory of matrix mechanics with Heisenberg and Jordan. They suffice to illuminate that in the context of great efforts to experimentally capture all aspects of atomic collisions, molecular formation and dissociation it was self-evident for Born that any theory of quantum mechanics must address the question of collision and scattering experiments.

Heisenberg, who once skipped the practical classes with Franck, which Born had suggested to take after his weak performance at his Munich doctoral examination in July 1923,[26] celebrated the breakdown of a space-time description of single objects in the atomic realm governed by matrix mechanics as a revelation, and even proclaimed that the notion of electron or particle trajectories had to be abandoned completely.[27]

---

[23]Cf. for details (Im 1996, 86f.) and (Lemmerich 2007, 125ff.).

[24]Scattered information about the practices of coordinated and collective work towards matrix mechanics like at Born's home, where, e.g., sessions of discussion and computation were interrupted by joint music-making, as mentioned in recollections, letters and anecdotes, have not yet been systematically analyzed.

[25] Thanks to the "History and Foundations of Quantum Physics" project at the Max Planck Institute for the History of Science, which ran from 2006–2012, a number of publications emerged that probably used an unprecedented broad basis of archival sources. Cp. e.g. Duncan and Janssen (2007), Joas and Lehner (2009), Blum et al. (2017), Jähnert (2019).

[26]AHQP Interview with James Franck, 9–14 July 1962.

[27]Cp. the reconstruction of Heisenberg's views and the confrontation with Schrödinger from a biographical perspective in Cassidy (1992, 214–216), the deconstruction of a matrix mechanics

Born, however, immediately started to seek a framework that would combine the new impossibility of accounting for all mechanical quantities at the atomic level with the experimental evidence constantly obtained from collision and scattering experiments in Franck's laboratory. For this reason, Born, Jordan and Nordheim still employed difference calculus to explain excitations of atoms in collisions even with matrix mechanics at hand, which, however, was not yet capable of dealing with this particular case. Only after Schrödinger's wave mechanics did this become possible (Born et al. 1925; Nordheim 1926). This example illustrates how the different lines of development did not all meet exactly in one single point.

Therefore, it appears rather straightforward that it was Born who proposed his statistical interpretation of quantum mechanics in June 1926 and stressed that this step would hardly have been feasible without his collaboration with Franck (Born 1926b). By formulating the scattering problem in terms of Schrödinger's wave mechanics, but interpreting the ψ-function as a new entity that allows to determine the probability of a quantum state (rather than that of a transition), solving the equations with appropriate approximations and expanding the result in terms of plane waves that represent free particles, Born was able to account for the probabilities of the respective scattering outcomes. As a tribute to Franck, he included inelastic scattering, thus providing a quantum mechanical description of the very Franck-Hertz experiment that is regarded as the first corroboration of a quantum atomic structure as introduced by Bohr (Born et al. 1926).

When Born traveled to the US at the end of 1925 shortly after the formulation of the Göttingen matrix mechanics and before Schrödinger published his technically more traditional wave mechanics—although leaving open what these waves would actually represent—he fully remained on this track of mastering collision processes. From this perspective, it is also clear that the often mentioned near miss of wave mechanics, when he and Norbert Wiener, who had just worked out a general operator calculus, introduced operator techniques for quantum theory, does not do justice to his approach (Wiener 1926). Since the relation between energy and time appeared most important for collision processes, while direct observation of the position seemed to be unfeasible, Born and Wiener merely replaced the energy with a differential operator $\frac{\hbar}{i}\frac{d}{dt}$. However, there was no motivation to relate linear momentum p to the coordinate derivative $\frac{d}{dx}$ (Born and Wiener 1926).

Born's self-understanding as a physicist with strong attention to experimentation, which had developed to some extent after the war, probably also overshadowed his identity as a theoretician and mathematician trained by Hilbert and Minkowski. In any case, the collaboration between mathematicians and physicists in Göttingen, which was so close when Hilbert turned into an ersatz physicist in the years before World War I, had somewhat loosened after the war when the experiment claimed some priority over theory.

---

revolutionary narrative in Beller (1983) and the suggestion of a dialogical historiography of quantum mechanics in Beller (1999, esp. Chap. 2), which, e.g., demonstrates that Heisenberg's call for the elimination of unobservables was an ex post facto statement.

Consequently, it comes as no surprise that Hilbert complained about a lack of interest on the part of mathematicians when so few from this field showed up to Heisenberg's presentation of matrix mechanics in Göttingen in September 1925, admonishing them not to ignore "great things that come into the world in Göttingen.[28] Due to Hilbert's illness around that time, Born had not yet heard his view on matrix mechanics, which had become thoroughly established by the three-man paper of November 1925 before he left for America (Born et al. 1926).[29] When Hilbert could finally take a closer look at quantum mechanics a year later, typically within a lecture on the mathematical foundations of quantum theory, wave mechanics and Born's statistical interpretation had arrived, so that now everything fell to its (mathematical) place (Hilbert 2009b, 605–707; Hilbert et al. 1927).

Since the four publications in *Zeitschrift für Physik* in the volumes 33, 34, 35 and 38 almost immediately codified the new quantum mechanics of Göttingen origin within just twelve month (Heisenberg 1925; Born and Jordan 1925a; Born et al. 1926; Born 1926a), and due to the fact that these papers can still be read today almost 100 years later as a reasonable introduction to the theory for physicists, because the new concepts of infinite-dimensional matrices, non-commuting (differential) operators, eigenvalue spectra and Hilberts spaces were an immediate and lasting success, there is not much that hints at a long and entangled development, in which the mathematical tools so much identified today as part of quantum mechanics were actually developed much earlier, without anticipating this end, but with the firm conviction that investing in such a kind of mathematics could revolutionize physics. But the story was long and entangled, and I have tried to demonstrate this in my study.

---

[28]Undated notes of a speech c. September 1925, Hilbert papers 657, 33.

[29]Born had asked in a letter to Hilbert, 28 November 1915, UAG 40A, Nr. 21, what he would say "regarding our attempt at a quantum mechanics."

Chapter 6
Appendix: Selected Documents

Document 1

From a letter from David Hilbert, on vacation in Alassio, to Arnold Sommerfeld in Munich, 5 April 1912. (DMA, HS 1977–28/A, 141).

Wie Sie wissen, haben wir bisher aus den Zinsen der Fermatstiftung die Kosten für die Einladungen der Herrn Poincaré und Lorentz bestritten. Wir möchten aber dieses Jahr, wo zu allem Sonstigen auch noch der internationale Kongress für Mathematik in England kommt und da diesmal auch Klein nicht dabei sein kann, nicht wieder in derselben Weise zu einer wissenschaftlichen Fermat-Woche einladen. Ich habe mir daher folgenden Ersatz in bescheideneren Dimensionen gedacht: Ich selbst lese dieses Semester Mittwoch u. Sonnabend 9–11 Uhr über Principien und Axiome der Physik. Wie wäre es, wenn für die beiden letzten Doppelstunden des Semesters also etwa 29. Juli u. 2. Aug. 9–11 Uhr statt meiner Sie eintreten würden? Diese Zeit würde den Göttinger Dozenten u. jüngeren Mathematikern u. Physikern wohl am besten passen, so dass ich für ein gut besuchtes Auditorium bürgen könnte. Der Gegenstand bliebe Ihnen ganz überlassen: Am besten wohl Strahlungsth[eorie] und Quantentheorie. Auswärtige würden wir nicht besonders einladen, wenngleich solche, wenn sie kommen, uns natürlich sehr willkommen sein würden. Ein gutes Honorar (ich denke etwa 1000 M.) könnte ich Ihnen aus der Fermatstiftung in Aussicht stellen. Wie denken Sie darüber? Es wäre auch eine gute Gelegenheit uns einmal wiederzusehen und ausführlich zu sprechen. Alle Göttinger Kollegen würden sich ausserordentlich freuen, am meisten aber ich.

Document 2

From a letter from Peter Debye after his appointment in Utrecht to Arnold Sommerfeld in Munich, 3 November 1912. (DMA, HS 1977–28/A, 61).

Mit meinem Colleg geht alles schon seinen gewohnten Gang, ich lese Mechanik, weil es nötig ist und Kinetische Theorie von Magnetismus und Dielectrica, weil es mir Spaß macht. Obwohl das hier nicht gebräuchlich ist, habe ich doch die Semestereinteilung beibehalten, damit nicht zu viele Dinge nebeneinander her laufen, wie das bei einer Jahreseinteilung nötig wäre. Das zweite Semester kann ich anfangen lassen, wann ich will und so die Zeit vernünftig ausnutzen. Für das zweite Semester habe ich angekündigt (1) Thermodynamik

© The Author(s), under exclusive license to Springer Nature Switzerland AG 2019
A. Schirrmacher, *Establishing Quantum Physics in Göttingen*,
SpringerBriefs in History of Science and Technology,
https://doi.org/10.1007/978-3-030-22727-2_6

(2) Nernst'sches Theorem und Quantentheorie. (...) Ich muß noch eine Bemerkung machen, damit Du ganz genau weißt, wie ich über die Berufung denke. Du hättest gerne gehabt, daß ich Utrecht ausgeschlagen hätte mit Rücksicht auf Leiden. Das schien mir nicht so sehr wünschenswert, denn der einzige Grund dafür wäre das Erreichen des ideellen Wertes gewesen: Nachfolger von Lorentz heißen zu können. Das ist nun aber doch schließlich nicht anders als rein äußerliche Sache zu betrachten, etwa wie ein Orden. Das was ich mache oder machen werde, würde dadurch in keiner Weise beeinflußt und das ist die Hauptsache, denn meine Stellung hier ist tatsächlich genau dieselbe, wie sie in Leiden gewesen wäre. Also kurz und gut ich bin zufrieden, wie es ist. Wenn Du nun frägst, wie es kommt, daß gerade Ehrenfest genommen wurde: nun das habe ich gut merken können, dafür hat Einstein gesorgt ... (...) In Deinem vorigen Briefe sprichst Du weiter davon, daß ich nach Göttingen zu der Physikerzusammenkunft kommen soll, von der ich aus Deinem Briefe zuerst höre und von der ich absolut nicht weiß, was sie sein soll oder wann sie abgehalten wird. Daß Du versucht hast mich mit dorthin zu kriegen, hat mich sehr gefreut und ich danke Dir wirklich sehr und gerne dafür. Du mußt es mir aber nicht übelnehmen, wenn ich nun so das Gefühl habe nicht als ungeladener Gast dort plötzlich auftauchen zu können. Du sagst zwar, Du könntest mich offiziell einladen, aber in Wirklichkeit ändert das doch nichts daran, daß die Herren denken werden, ich hätte bei etwas mehr Taktgefühl empfinden müssen, daß ich besser wegbleiben könne. Ich möchte nicht den Eindringer spielen, der in die Küche gehört, aber nun mal mit in den Salon hineingekommen ist und dann dort allerdings recht freundlich geduldet wird. Sei mir deshalb nicht böse, ich möchte Dir gerne, ohne das fürchten zu müssen, offenherzig sagen können, was ich denke. (...)

Document 3

Application from the Göttingen Königliche Gesellschaft der Wissenschaften to the Prussian Ministry of Culture regarding annual funds of 5000 Marks for a guest professorship in the mathematical sciences, 7 July 1913. (GStA PK. 76 V a, Sekt. 6, Tit. IV, Nr. 1, Bd. XXVII, Bl. 153–155, duplicate in UA-Gö Hilbert Papers 494, Nr. 7, Bl. 13–15).

Die persönliche Zusammenarbeit in der soeben charakterisierten Weise erscheint uns im gegenwärtigen Zeitpunkte besonders wichtig und fruchtbar für die Disziplinen der mathematischen Erkenntnistheorie einerseits und der theoretischen Physik und Astronomie andererseits. Die beantragte Institution soll daher außer der reinen Mathematik besonders der Pflege dieser Disziplinen zugute kommen, indem wir alljährlich für das Sommersemester einen Gelehrten als Gast hierher nach Göttingen berufen, der entweder in hervorragender Weise einen Spezialzweig der reinen Mathematik vertritt oder auf einem der genannten Nachbargebiete eine führende Rolle einnimmt. Die Aufgabe der Berufenen besteht darin, an den gelehrten Vereinigungen insbesondere der mathematischen und physikalischen Gesellschaft, wie sie an unserer Universität bestehen, tätig teilzunehmen, mit den hiesigen Dozenten in persönliche Beziehung und wissenschaftlichen Ideenaustausch zu treten und endlich einen Cyclus von Vorlesungen aus seinem Spezialgebiet abzuhalten, welche für die Dozenten, Assistenten, Doktoren, und evtl. einer in bestimmter Masse auszuwählenden Studenten in höheren Semestern berechnet sind. Demnach würde die in Rede stehende Institution nur akademisch wissenschaftlichen Interessen dienen und keinen unmittelbaren Einfluss auf den allgemeinen Unterrichtsbetrieb der Universität ausüben. Die von uns beantragte Institution wäre eine Förderung der theoretischen Forschung auf rein akademischen Boden wie eine solche den experimentellen Wissenschaften in der letzten Zeit so mannigfach zuteil geworden ist, so insbesondere neuerdings durch die Forschungsinstitute der Kaiser-Wilhelm-Stiftung. Im Vergleich mit den für die experimentellen Wissenschaften aufgewandten Mittel erscheinen die hier beantragten Mittel als ausserordentlich bescheiden. Wir erwähnen noch, daß die genannten, der Mathematik benachbarten Gebiete – mathematische Erkenntnistheorie, theoretische Physik und mathematische Astronomie – sämtlich auf den deutschen

Universitäten nicht so vielseitig vertreten sind ... als notwendig erscheint. Dieser Übelstand ist im Augenblick vor allem für die theoretische Physik fühlbar; denn da das Ausland in dieser Hinsicht durchweg besser gestellt ist, so haben wir häufig unsere tüchtigsten, jungen theoretischen Physiker an das Ausland abgeben müssen... Es muß betont werden, daß die Pflege der theoretischen Seite der Naturwissenschaften und die Kultur der in ihr wirksamen reinen Geistesarbeit ebenso dringende Notwendigkeit und Pflicht wie die Erforschung der experimentellen Tatsachen in der Naturwissenschaft ist.

Document 4

Addition to the memorandum of David Hilbert to the Prussian Ministry of Culture regarding the application for a guest professorship, circa 7 July 1913. (GStA PK. 76 V a, Sekt. 6, Tit. IV, Nr. 1, Bd. XXVII, Bl. 156–157, duplicate in UA-Gö Hilbert Papers 494 , Nr. 8, Bl. 19–20).

Zur eingehenden Motivierung der für die Göttinger Gastprofessur beantragten 5000 M. gestatte ich mir noch Folgendes zu bemerken: 1. Für die Berufung von Männern ersten Ranges aus dem Auslande wie H. A. Lorentz (Leiden), den wir vor allem wünschen, Rutherford (Cambridge), Hadamard (Paris) etc. glauben wir aus Gehalt für das Semester die volle Summe von 5000 M. nötig zu haben. 2. Wenn Inländer, die für das Sommersemester hierher nach Göttingen beurlaubt werden, oder jüngere Gelehrte berufen werden, so wird allerdings eine geringere Summe als Gehalt genügen. Es st aber zu bedenken, dass für die Gasprofessur - abgesehen davon, dass wir uns doch vorbehalten müssen gelegentlich auch einen reinen Mathematiker zu berufen – nicht weniger als drei grosse und gleich wichtige Nachbardisziplinen nämlich: mathematische Logik und Philosophie, theoretische Physik, und mathematische Astronomie in Frage kommen. Wenn daher die beabsichtigte neue Institution der Gastprofessur eine nachhaltige Wirkung erzielen soll, so wäre die gleichzeitige Berufung von mehreren Gelehrten nötig. Dies ließe sich nur erreichen, wenn unserem Antrage auf Bewilligung von 5000. M. entsprochen wird. Da nach dem Weggang von Brendel und Schwarzschild die mathematische Astronomie hier garnicht vertreten ist, so sind wir mit Professor Haar aus Clausenburg in Verbindung getreten und es ist Aussicht vorhanden, dass Haar für das Sommersemester 1914 Urlaub bekommt, uns hier in Göttingen mathematisch-astronomische Vorlesungen zu halten. Haar soll aus dem Fermatfonds honoriert werden und wenn dieser auch schon für verschiedene andere Zwecke in Anspruch genommen ist, so könnten wir ihn doch gelegentlich mit heranziehen um im Verein mit den beantragten 5000 M. die gleichzeitige Berufung mehrerer Gastprofessuren zu ermöglichen. 3. Überhaupt betrachte ich die von der Gesellschaft der Wissenschaften beantragten 5000 M. nur als einen Grundstock zur Förderung der theoretischen Forschungsgebiete; ich hoffe, dass, wenn wir erst die genannte Summe im Etat der Gesellschaft haben, es möglich sein wird, Stiftungen von privater Seite zu gewinnen – hat doch die Verwendung der Fermatgelder sich so gut bewährt, und den weitesten Anklang gefunden, so insbesondere der im Frühjahr 1913 einberufene Kongress über die kinetische Theorie der Materie mit den Vorträgen von Planck, Nernst , Debye, H. A. Lorentz, Sommerfeld, von Smoluchowsky, die je 800 Mark aus dem Fermatfont erhielten.

Document 5

From a letter from David Hilbert to Hugo Krüss[1] in the Prussian Ministry of Culture in Berlin concerning the establishment of a guest professorship, 3 October 1913. (GStA PK. 76 V a, Sekt. 6, Tit. IV, Nr. 1, Bd. XXVII, Bl. 158–162, duplicate in UA-Gö Hilbert Papers 494, Nr. 9, Bl. 23–27).

[1] Krüss (1879–1945), who graduated with a dissertation in physics in 1903, became an important employee in the Ministry and was honored with the title of professor in 1909.

Nach der allgemeinen Motivierung die die hiesige Kgl. Gesellschaft auf einstimmigen Beschluß in ihrer Eingabe an das hohe Ministerium gegeben hat und nach den speziellen Ausführungen, die ich in dem neulich an Sie gerichteten Briefe zur Begründung der Höhe gemacht habe, und die wesentlich die für die Berufung in Frage kommenden Persönlichkeiten betraf, glaube auch ich, wie Sie, die Sache am besten zu stützen, wenn ich noch in sachlicher Hinsicht eingehender ausführe, warum gerade die sich gegenwärtig auf mathematischem Fundamente vollziehende Umwandlung der physikalischen Grundbegriffe die innere Notwendigkeit der Schaffung einer solchen Institution, wie wir sie wünschen, in sich birgt.

Wenn ich in einem fahrenden Eisenbahnzuge in der Fahrtrichtung vorwärts gehe, so bewege ich mich gegenüber dem Erdboden... (...) Der Einsteinsche Gedanke ist seit David Humes Kritik des Causalitätsbegriffes die kühnste und gewaltigste wissenschaftliche Idee geworden. (...) In ihr [der neuen vierdimensionalen Weltmechanik] erscheinen schon früher entdeckte Tatsachen in neuem und einfacherem Zusammenhange so z.B. der von Kaufmann experimentell und von Abraham mathematisch erbrachte Beweis, dafür dass die Trägheit, diese auch dem Laien bekannte alltägliche Fundamentaleigenschaft der Materie ein rein elektromagnetisches Phänomen ist. (...)

Sie erkennen aus diesen kurzen Darlegungen, wie eng Mathematik und Physik gegenwärtig miteinander verknüpft und wie sehr dies beiden Wissenschaften aufeinander angewiesen sind. Während bisher die Mathematik wohl gelegentlich einzelne Probleme aus der Physik aufgriff und dann für sich allein in abstrakt mathematischem Sinne weiter behandelte und andererseits die Physik meist von der Mathematik nur technische Rechenregeln oder formale Rechenmethoden verlangte, verschmelzen hier beide Wissenschaften zu einem einzigen Wissensgebiete: es folgt daraus, dass der Mathematiker weit tiefer als bisher in das innerste Wesen der physikalischen Wissenschaft eindringen muss und ebenso der Physiker nicht mehr blos[s] mathematischer Laie bleiben darf. Die Schwierigkeiten, diesen Anforderungen zu genügen, sind für die Vertreter beider Wissenschaften an sich sehr bedeutende, durch blosses von Abhandlungen lassen sie sich gegenwärtig überhaupt nicht überwinden: denn der Physiker überspringt in der schriftlichen Darstellung leicht wichtige logische Schlußreihen, die er wegen der ihm geläufigen und anschaulich gegenwärtigen Experimente als selbstverständlich hinnimmt, wo für den Mathematiker oft gerade der Schlüssel zum Verständnis der physikalischen Vorgänge ruht. Dem Physiker andererseits ist es meist ganz unmöglich, dem abstrakten Gange einer modernen mathematischen Abhandlung zu folgen, selbst wenn diese ein ihm geläufiges Wissensgebiet betrifft.

Die populär gehaltenen vielgelesenen Schriften von H. Poincaré und E. Picard liefern fortlaufende Illustrationen dafür, dass die wichtigsten Fortschritte in den Naturwissenschaften durch die Verbindung der Grenzgebiete erreicht werden. "Gewisse Arbeiten" so sagt E. Picard in der Einleitung zu seiner Schrift über das Wissen der Gegenwart, "können nur durch das Zusammenwirken eines Mathematikers mit einem Physiker zur Durchführung kommen." Und wenn ich Ihnen eine solche spezielle nur gemeinsam von Mathematikern und Physikern zu lösende Aufgabe nennen soll, so ist es die Ergründung der Struktur des Atoms: ein großzügiges Problem, bis vor kurzem noch fast unzugänglich, in das gegenwärtig alle wissenschaftlichen Gedankenfäden zu münden scheinen, wenn auch seine definitive Lösung wohl noch in weiter Ferne steht! Die bedeutsamen Arbeiten, über die Struktur der Materie, die gegenwärtig besonders in England und Holland erscheinen, durch persönliche Einwirkung und wechselweiser Aussprache unserem Ideenkreis vertraut zu machen, wär eine der Aufgaben, die dem theoretisch physikalischen Gastprofessor für die nächste Zukunft zufiele und es wäre damit eine klaffende Lücke in unserem gegenwärtigen Wissenschaftsbetriebe ausgefüllt. Ähnlich, wenn auch nicht so augenfällig für die Fernstehenden - steht es mit den Beziehungen zwischen Mathematik und Erkenntnistheorie: ...

Document 6

From the protocol of the Tagung der Göttinger Vereinigung zur Förderung der angewandten Physik und Mathematik, held November 21–22, 1913 in Göttingen. Voigt requests support for spectroscopic equipment for 1200 Marks. Subsequently he speaks about the second Solvay Congress which he had attended. (GStA PK. 76 V a, Sekt. 6, Tit. X, Nr. 4 Adbih. Heft VII, pp. 45–47).

Der Herr Kollege Riecke und ich haben einen Bericht über die allgemeine Tätigkeit unserer Institute nicht eingereicht, weil letztere nicht eigentlich in das Arbeitsgebiet der G[öttinger] V[ereinigung] fallen. Dagegen habe ich mir erlaubt, ganz unkommentmässig in elfster Stunde noch ein Gesuch um eine Beihülfe für das von mir geleitete Institut einzusenden. Ich bin dazu gekommen einmal durch den Eindruck der in diesem Jahr ganz bedrohlichen an das Institut gestellten Anforderungen, die auf dessen starker Frequenz beruhen, sodann durch die Erwägung, dass die G.V. durch das Nichtzustandekommen des Juristenkurses in diesem Jahre einige Mittel besitzt, auf die sie nicht gerechnet hatte. Ich war der Meinung, dass man der letzteren an sich unerfreulichen Tatsache vielleicht eine für mein Institut erfreuliche Seite abgewinnen könnte. Das von mir geleitete Institut ist ein theoretisch physikalisches in dem Sinne, dass die Theorie darin herrschen ist; aber es ist nichtsdestoweniger ein Institut experimenteller Arbeit. Nur beschäftigen wir uns nicht mit der Pionierarbeit in neuen, unaufgeklärten Gebieten, sondern verfolgen Aufgaben, die mit der Prüfung und Ausgestaltung der Theorien in bereits erschlossenen Gebieten zusammenhängen, und Bestimmten speziell die Zahlenwerte physikalischer Konstanten, die in den Theorien auftreten. Das ist eine nach aussen einigermassen unscheinbare Arbeit, die aber nichtsdestoweniger notwendig und nützlich ist. Diese Tätigkeit steht auch, wie ich bei einer früheren Gelegenheit ausgeführt habe, in engem Zusammenhang mit dem höheren wissenschaftlichen Unterricht in der Physik. Die Göttinger Vereinigung hat meinem Institut, trotzdem dasselbe nicht zu ihrem eigentlichen Unterstützungsgebiet gehört, wiederholt ihre willkommene Hilfe angedeihen lassen. Zum letzten Male vor zwei Jahren, wo sie zusammen mit der Unterrichtsverwaltung die Mittel zur Anschaffung eines modernen erstklassigen Gitterspektoskopes gespendet hat.

Document 7

From a letter from David Hilbert to Oberregierungsrat Ludwig Elster in the Prussian Ministry of Culture, 6 April. 1914. (GStA PK. 76, Nr. 591, Bl. 210).

Für das bevorstehende Etatsjahr habe ich meinen bisherigen 2. Assistenten Dr. Landé (theor. Physik) als ersten Assistenten engagiert, der mit für die Vorbereitung meiner theoretisch physikalischen Vorlesungen unentbehrlich geworden ist und da mir die Erhöhung seiner Renumeration von 800 auf 1200 Mark notwendig erschien. Mein bisheriger erster Assistent Dr. Hecke, Privatdozent für Mathematik wird mir seine Assistentendienste noch ein weiteres Jahr widmen und soll daher mein zweiter Assistent werden. Da nun Dr. Hecke bisher 1200 M. erhalten hat und mir unschätzbare Dienste leistet, so beantrage ich, ihm ausnahmsweise für das kommende Etatsjahr statt 800 M, die sonst mein zweiter Assistent bekommt, 1200, bewilligen zu wollen. Ich gestatte mir noch zu bemerken, dass ich im kommenden Sommersemester meine Asisten[ten] in besonderem Grade in Anspruch nehmen muss: es ist nämlich teils aus der Wolfskehlstiftung, teils aus dem vom Ministerium und der Kgl. Gesellschaft der Wissenschaften bewilligten Mitteln zwei "Gastprofessoren" nämlich Debey-Utrecht für theoretische Physik und Haar-Klausenburg für theoretische Astronomie zu engagieren gelungen.

Document 8

From a letter from Voigt to Otto Naumann, director in the Prussian Ministry of Culture, 20 May 1914 (GStA PK. 76 V a, Sekt. 6, Tit. IV, Nr. 1, Bd. XXIV, Bl. 71–73).

In dieser Eingabe ist auf meine Veranlassung ausgeführt, daß es angemessen erscheint, dem in die betreffende Stelle zu Berufenden (neben dem persönlichen Ordinariat) die Leitung der bisher mir unterstehenden Abteilung B des physikalischen Instituts zu übergeben. Es ist dies nicht nur deshalb geschehen, weil dem der Fakultät in erster Linie erwünschten Kandidaten Debye von Zürich und nun Frankfurt (neben dem Ordinariat) ein nicht unbedeutendes selbständiges Institut angeboten wird und Göttingen ihn unter anderen Bedingungen nicht gewinnen könnte, – wir halten diese Anordnung auch für an und für sich zweckmäßig. Eine starke Schaffenskraft, wie sie bei Debye unzweifelhaft vorliegt braucht für ihre Betätigung freien Raum, und so sehr ich erwarte mit dem neuen Kollegen in wissenschaftlicher Arbeit auch im Institut zu kooperieren, so würde er doch mein verbleiben an der Spitze des Instituts naturgemäß als Einengung empfinden. Bezüglich der Wünsche, die der neue Institutsdirektor geltend machen wird, beehre ich mich folgendes auszuführen: der Universität ein theoretisch-physikalisches Institut zu schaffen, das im Sinne des Ideales meines Lehrers Franz Neumann die breiteste Berührung zwischen Theorie und Beobachtung vermittelt. Ein solches Institut kann naturgemäß nicht an jeder Universität bestehen ... (...) Es folgt hieraus, daß man in dem mit Mühe und mit Opfern auf eine gewisse Höhe gebrachten Institut der Betrieb ferner in der bisherigen Weise aufrecht erhalten werden soll, ähnliche Geldmittel auch weiterhin zur Verfügung stellen müssen. Kaum wird ein neuer Direktor in der Lage und bereit sein, ähnliche Zuschüsse zu leisten, wie ich aus Interesse an der Begründung und Hebung des Institutes geleistet habe. Um das Institut nicht Not leiden zu lassen, wird also der Etat für sachliche Geldmittel ... einer nachdrücklichen Aufbesserung bedürfen.

Document 9

From a letter from Dean Gustav Körte of the philosophical faculty at Göttingen to the Prussian Ministry of Culture, 28 May 1914. This memorandum was presented in the session of the philosophical faculty on 28 May 1914 after Voigt had reported on the prehistory of the motion for a replacement professorship in theoretical physics. The proposed text was unanimously authorized.(GStA PK. 76 V a, Sekt. 6, Tit. IV, Nr. 1, Bd. XXIV, Bl. 74–76; also Göttingen UA 4 I 105, Bl. 149–151).

Eine Hochschule, die in theoretischer Physik vollwertig vertreten sein will, muss diese wichtige neue Richtung in Lehre und Forschung ausgiebig berücksichtigen, eventuell durch Heranziehung neuer Mitarbeiter. Zur Illustration der hierdurch entstandenen Bewegung sei darauf hingewiesen, dass das ganz allgemein und auch gerade bezüglich jüngerer Kräfte so reich ausgestattete Berlin in allerletzter Zeit zwei der bedeutendsten Vertreter der neuen Richtung gewonnen hat, resp. zu gewinnen sucht: Einstein in eine akademische, Laue in eine Universitätsstellung (...) Der gegenwärtige Inhaber der Professur für theoretische Physik, der Kollege Voigt, hat uns nun erklärt, dass er in seinem Alter den neuen bedeutungsvollen Entwicklungen wirklich gerecht zu werden nicht imstande ist, zumal seine gepflegten Arbeitsgebiete nach anderer Seite hin liegen. (...) Derjenige, dessen Gewinnung uns in allererster Linie erwünscht sein würde, der gegenwärtig in Göttingen als Gastprofessor wirkende Herr Debye aus Utrecht, bewährt z.B. diese Eigenschaft [eines Ordinariats würdig zu sein] dadurch, dass er nicht nur in Utrecht eine ordentliche Professur inne hat, sondern auch augenblicklich von Frankfurt und von Zürich in Ordinariate begehrt wird. Wir haben Ursache zu hoffen, dass Herr Debye das Extraordinariat in Göttingen, verbunden mit einem persönlichen Ordinariat cum iure succedendi sowohl der gegenwärtig innegehaltenen, als auch den beiden angebotenen Stellen vorziehen würde. Könnten wir ihm aber nicht eben jetzt eine Berufung nach Göttingen in Aussicht stellen, so würde er unserer Hochschule, unzweifelhaft in bedauerlicher Weise, definitiv verloren gehen.

Document 10

From a draft of a letter from Hilbert to Geheimrat Ludwig Elster in the Prussian Ministry of Culture, 25 January 1915 (UA-Gö Hilbert Papers, folder 466, Nr. (1). Here with markings of crossed out and added words (underlined).

> Während der weltumwälzenden ~~großen~~ Geschehnisse draußen spielt sich hier hier eine Entwicklung ab, die für den engen Kreis der math.-naturwissenschaftlichen Kreis in Verhältnissen von nicht geringer Bedeutung sind und da Sie, hochver. Herr Geheimrat, indem Sie die Herberuf[ung] Debyes ~~ermöglichten~~ herbeiführten, der Urheber der Entwicklungen sind, so kann ich nicht umhin, ~~in aller Kürze~~ in Ihnen das[?] mitzuteilen. Durch den Beitritt[?] Debyes hat das math. Seminar hier ein solche[?] Höhe err[ungen?], daß fast alle math[ematisch]-phys[ikal]ischen Lehrkräfte daran teilnehmen und die Dispute[?] dann gestalten sich zu wissens[chaftlichen] Taten. Debye ~~ist~~ erweist sich [als] der Newton der ~~Chemie~~ Molekül Physik und wir haben jetzt insbes[ondere] durch seine neuesten Entdeckungen um Weihnachten [1914] herum die solange vergeblich gesuchte in weiterer Ferne geglaubte Grundlage einer neuen math[ematischen] Chemie. So ist zugleich für mich persönlich in wissenschaftlicher Hinsicht Debye wirklicher Ersatz für Minkowski geworden. (...) Übrigens hat Debye für morgen die Vorladung zur militär-ärztlichen Untersuchung erhalten... Nun fehlt mir nur noch Willy Wien, dessen Herberufung wie ich glaube nicht sehr sicher wäre evtl. nach Friedensschluß!

Document 11

From a letter from Wilhelm Wien to Otto Naumann, director in the Prussian Ministry of Culture, 4 December 1915 (GStA PK. 76 V a, Sekt. 6, Tit. IV, Nr. 1, Bd. XXIV, Bl. 341–342).

> Ich nehme mir die Freiheit an Ew. Excellenz ein Schreiben in einer Frage zu richten, welche für die ganze Physik von erheblichem Interesse ist nämlich die Besetzung des Ordinariats für experimentelle Physik in Göttingen. So ungern ich mich ohne unmittelbare Aufforderung in Berufungsfragen einmische, ... so scheint es mir doch diesmal im allgemeinen Interesse unserer Wissenschaft geboten zu sein. ... Es ist eine Stelle, die für die künftige Entwicklung der deutschen Physik sehr bedeutsam werden kann..." oder zu "großer Bedeutungslosigkeit" herabsinken würde. (...) Wenn nach Göttingen nicht ein wirklich produktiver Kopf kommt, so werden vor allem die vielen jüngeren Arbeitskräfte die sich dort wie kaum an einer anderen Universität sammeln, nicht in richtiger Weise gelenkt werden und es werden besonders die vielen Anregungen, die dort wie nir[gends] wo anders von mathematischer und theoretisch-physikalischer Seite ... ungenutzt bleiben. Naturgemäß werden auch die Rückwirkungen fehlen, welche von Seiten der experimentellen Physik auf die Theorie und die Mathematik stattfinden sollen und die in Göttingen besonders wünschenswert sind. Denn bei der dortigen Entwicklung überwiegen gegenwärtig die mathematischen die experimentellen so erheblich, daß sie, man kann sagen, gegen ihren Willen fast alleinherrschend geworden sind. Ich glaube, daß die Mathematiker selbst darunter leiden und es gern sehen würden, wenn die experimentelle Physik einen größeren Einfluß auf das wissenschaftliche Leben Göttingens gewinnen könnte. Nach meiner Meinung das Schlimmste wäre die Berufung einer jungen physikalischen Mittelmäßigkeit, wie sie nach dem von mir gewonnenen Eindruck einzutreten droht. Dann wäre die Göttinger Physik auf mehr als ein Menschenalter lahmgelegt... Nach meiner Überzeugung würde dies eintreten, wenn einer der von manchen Seiten jetzt besonders empfohlenen Herren Franck, Pohl, Edgar Meyer nach Göttingen berufen werden. (...)

... ich kann nur bedauern, daß die Berufung Starks, den ich trotz seiner persönlichen Mängel für die bei weitem geeignetste Persönlichkeit halte, aus persönlichen Gründen unmöglich erscheint. Aber wenn die Berufung Starks wirklich ausgeschlossen ist, wäre es immer noch besser, einen nicht mehr ganz jungen Physiker zu nehmen, der jüngeren Forschern Arbeitsgelegenheit verschafft, als einen in organisatorischen Fragen ganz unerfahrenen zu berufen, der schließlich für vier Jahrzehnte dem Göttinger Institut voraussichtlich den Stempel der Mittelmäßigkeit aufdrücken würde.

Document 12

From a letter from David Hilbert to Otto Naumann, director in the Prussian Ministry of Culture, 24 December 1915 (UA-Gö Hilbert Papers, folder 466, Nr. 2).

Vom Briefe W. Wiens an Eure Excellenz in Angelegenheit der Wiederbesetzung des hiesigen Ordinariats für Experimentalphysik hat mir Herr Kollege Voigt Mitteilung gemacht. Meine Einwendungen gegen Stark-Aachen beziehen sich nur auf die Göttinger Stelle, für die ich allerdings nach wie vor die Berufung jedes anderen Physikers außer StarkPhysikers außer StarkPhysikers außer Stark, auch eines solchen, der nicht auf der Liste steht, erheblich vorziehe. Doch sehe ich die Gefahr jetzt nach unserem Vorschlag und der neuerlichen Erklärung Eurer Excellenz als beseitigt an. Dem sehr gefälligen Wunsch Euer Excellenz über die neuen Untersuchungen von Debye näheres zu erfahren, entspreche ich außerordentlich gerne. Das Ziel desselben ist die experimentelle Erforschung der Materie bis in das Innere des Einzelatoms hinein. Die zur Erreichung dieses Ziels verwandte Versuchsanordnung habe ich in meinem neulichen Briefe als ein höchst verfeinertes Mikroskop bezeichnet. Damit hat es folgende Bewandtnis: ... (...) ... es wird die von chemischer Seite schon als Hypothese angenommene Ringanordnung der sechs Kohlenstoffatome festgestellt und der Abstand zweier Atome voneinander zu 0,000 000 62 m.m. gemessen. Es liegt schon nach diesen ersten Proben auf der Hand, daß damit ein unabsehbares neues Feld der Experimentalforschung eröffnet ist, und ich möchte – mit etwas Kühnheit – sagen, daß mir eigentlich von nun an keine auf die Struktur der Materie bezügliche Frage mehr unbeantwortet bleiben kann, so groß auch noch die bis zu dem Ziel notwendig zu leistende Arbeit sein mag.

Document 13

From a letter from Otto Naumann, director in the Prussian Ministry of Culture, to Wilhem Wien, 15 January 1916 (GStA PK. 76 V a, Sekt. 6, Tit. IV, Nr. 1, Bd. XXIV, Bl. 366–367).

Obgleich manche persönliche Eigenschaften an ihm seine Berufung an eine Hochschule nicht leicht machten, habe ich ihn seinerzeit dem Herren Minister für die zweite Physikerstelle in Hannover und später für die erste Physikerstelle in Aachen empfohlen. ... immer wieder durch Herrn Professor Rubens veranlaßt worden ... Freilich auf der anderen Seite erklärte er stets, an sein Institut würde er Professor Stark keinesfalls nehmen. Gegen die Berufung nach Göttingen sprachen sich dann sehr energisch die nächstbeteiligten Göttinger Professoren aus, die zu diesem Zwecke hier persönlich vorstellig wurden. (...) Sieht man von diesen jüngeren Vertretern ab [Meyer, Pohl, Koch], so frage ich: wer ist denn sonst der geeignete Experimentalphysiker für Göttingen? (...) ... auch folgende Lösung in Frage gekommen: Professor Debye in Göttingen, der zugleich anerkannter tüchtiger Experimentalphysiker ist, wird als Ordinarius für Experimentalphysik an die Spitze der physikalischen Institute in Göttingen gestellt, und die von ihm bekleidete Professur – ein Extraordinariat – wird einem zweiten jüngeren Physiker, z.B. Dr. Pohl, übertragen, etwa in der Weise, daß beide Herren abwechselnd Experimentalphysik lesen. Die theoretische Physik könnte dann noch auf Jahre hinaus durch Professor Voigt in Vorlesungen vertreten werden.

Document 14

From a letter from Wilhelm Wien to Otto Naumann, director in the Prussian Ministry of Culture, 17 January 1916 (GStA PK. 76 V a, Sekt. 6, Tit. IV, Nr. 1, Bd. XXIV, Bl. 368–369).

Zur ersten Liste möchte ich, obwohl sie als erledigt anzusehen ist, das folgende äußern. Falls eine Berufung ergangen wäre, würde ich sie ernstlich in Erwägung gezogen haben und zwar ohne Rücksicht auf die Möglichkeit einer Berufung nach München. Eine solche ist nicht aktuell. Niemand weiß, wann Herr Röntgen zurücktritt und wir haben Beispiele genug, daß die Universitätslehrer von 80 Jahren u. mehr im Amte bleiben. Allerdings kann ich nicht sagen, ob Verhandlungen mit mir zu Ziele geführt hätten, weil in der That die Verhältnisse des Göttinger Institutes nicht günstig sind. Die der Experimental-Physik zugewiesenen Räume sind verhältnismäßig klein, da die theoretische Physik sehr viel belegt hat. Wenn die Göttinger Kollegen sich so sehr gegen Stark verwahrt haben, so findet das auch in den eigentümlichen Institutsverhältnissen eine gewisse Berechtigung. Ein Zusammenarbeiten zwischen Debye u. Stark unter diesen Umständen würde so gut wie ausgeschlossen ..[?]. Die Sache wäre nur gegangen, wenn man Stark das ganze Institut überlassen und Debye anderswo untergebracht hätte. (...) Jeder der drei vorgeschlagenen würde die Experimentalphysik neben dem unvergleichlich bedeutenderen Debye kaum ausreichend vertreten können. Wenn ich von älteren Physikern gesprochen habe, die meiner Meinung nach vorzuziehen wären, so hatte ich dabei Zenneck ..[?] Dißelhorst (Braunschweig) im Auge... Beide sind keine großen Bahnbrecher... Zenneck ist ein vorzüglicher Organisator... (...) Falls Debye geneigt ist, die Professur für Experimentalphysik zu übernehmen, müßte er es meiner Meinung nach ganz thun ... empfehlen einen anderen Theoretiker zu berufen... Mie in Greifswald oder Laue in Frankfurt. Mie ist ein ganz hervorragender Theoretiker, was die Göttinger selbst am besten wissen. Debye und Mie würden jedenfalls die Physik so glänzend vertreten, daß ich ... diese Lösung ... für die bei weitem beste halte."

Document 15

From a letter from David Hilbert and Felix Klein to the Prussian Ministry of Culture, 29 January 1916 (GStA PK. 76 V a, Sekt. 6, Tit. IV, Nr. 1, Bd. XXIV, Bl. 372–373).

Von dem gleichen Wunsche aus erlauben wir uns hier auch noch über Debye's augenblickliche finanzielle Stellung einiges anzugeben. Als Debye im Sommer 1914 nach Zürich in die gleiche Stellung berufen wurde, in die jetzt Edgar Mever eintritt, wurde ihm dort eine Jahreseinnahme von 25.000 Frs. = 20.000 Mark in sichere Aussicht gestellt. Aber Herr Geheimrat Elster war nicht in der Lage, ihm in Göttingen eine höhere Summe als 12.000 M. zu garantieren. Daher haben die Göttinger Fachkollegen Debyes von sich aus damals Maßregeln getroffen, um Prof. Debye, dessen Verbleiben allseitig als eines der wichtigsten Interessen der Universität erachtet wurde, wenigstens die Hälfte der Spanne zwischen 12.000 M. und 20.000 M. zu ersetzen. Es ist dies, wie Ew. Exzellenz damals vielleicht nicht ausdrücklich mitgeteilt wurde, in der Weise gelungen, dass a) Prof. Voigt zugunsten von Debye auf die Hälfte seines Gehalts verzichtet (was 2000 M. ergab), und b) die Fermatkommission der Gött. Ges. d. Wissenschaften einwilligte, aus den Zinsen des Wolfskehlschen Legats bis auf weiteres 2000 M. zuzuschiessen. Diese 4000 M. jährlich sollten nun wohl bei jetziger Gelegenheit seitens des Staates irgendwie abgelöst werden. In der Tat kann doch das Provisorium nicht dauernd aufrecht erhalten werden. ... Denn Prof. Debye erreicht jetzt schon, in Kriegszeiten, wo er nur eine theoretische Vorlesung hält, ungefähr die Höhe meiner Kolleggeldgarantie und seine Einnahme aus Kolleggeldern muss doch selbstverständlich bedeutend wachsen, wenn er seinen Teil der Experimentalphysik mitübernimmt.

Document 16

From a letter from the Minister of Culture (Konrad Haenisch) to Minister of Finance (Friedrich Erzberger), 9 January 1920, discussing the national dimension of Debye's imminent departure from Germany as well as the danger that Hilbert also might leave Göttingen (GStA PK. 76 V a, Sekt. 6, Tit. IV, Nr. 1, Bd. XXVI, Bl. 166–168).

... [wird] die Gefahr immer drohender, dass wir Deutschland, nachdem wir es in immer steigendem Masse von materiellen Gütern entblösst worden sind ist, nun auch an geistigen Werten verarmen. Wie auf wirtschaftlichem Gebiet benutzt das Ausland den Tiefstand der Valuta, um besonders ausgezeichnete Vertreter deutscher Wissenschaft für sich zu Bedingungen zu gewinnen, denen die preussische Unterrichtsverwaltung auch nicht annähernd folgen könnte, wenn sie sich an die für normale Verhältnisse gegebenen ... Richtlinien halten müsste. ... Pflicht ..., das Aeusserste zu tun, um die drohende geistige Verarmung und die daraus folgende Ausschaltung Deutschland auch aus dem geistigen Wettbewerb der Nationen zu verhindern. Vielmehr muss sie bestrebt sein, im Rahmen des Möglichen über die heimatlichen Verhältnisse im engeren Sinne hinaus nationale Kulturpolitik zu treiben, um das Ansehen der deutschen Wissenschaft in der Welt zu erhalten ... ...

Von dem Polytechnikum in Zürich hat der Göttinger Physiker Debye einen materiell wie ideell sehr vorteilhaften Ruf erhalten. ... dessen Zugehörigkeit zur deutschen Wissenschaft immer ein besonderer Gewinn für Deutschland gewesen ist, ... Der Versuch, ihn abzuhalten, muss schon aus Gründen des nationalen Ansehens gemacht werden. Hierfür ist vor einiger Zeit auch die Vertretung der gesamten deutschen Studentenschaft entscheiden eingetreten, und auch die Preussische Landesversammlung hat in einer förmlichen Anfrage den Wunsch, Debye für Deutschland zu erhalten, ausdrücklich kundgegeben. Zudem ist zu befürchten, dass, wenn Debye nach Zürich geht, ihm der berühmte Göttinger Mathematiker Hilbert folgt und dass so die in ihrer Bedeutung weltbekannte naturwissenschaftliche Fakultät der Universität Göttingen ihres Gehalts im wesentlichen beraubt wird.

Document 17

From a Letter from Otto Wallach to Arnold Sommerfeld, 21 February 1920 (DMA Sommerfeld papers, NL 89, box 019, folder 5, 7).

Begreiflicherweise besteht der Wunsch, den grossen Verlust, den wird durch Debye's Fortgang erleiden, wett zu machen. Da ist natürlich, dass sich der Blick auf Sie richtet – aber, das kann doch nur mit einem gewissen Zagen geschehen, denn es herrscht die Empfindung vor, dass Sie für uns wahrscheinlich ganz unerreichbar sind. [...] Sollten Sie uns Ihre Bereitwilligkeit unzweideutig zuerkennen geben wollen, einem Ruf hierher Folge zu leisten, so würden Sie Zweifels nicht nur an erster, sondern an einziger Stelle der Regierung als zu Berufender vorgeschlagen werden und Ihren Entschluss würde man hier allgemein mit grösstem Jubel begrüssen. Besteht aber keine Aussicht, Sie hierher zeihen zu können, so müssen wir eben resignieren.

Document 18

From a letter from the Göttingen Dean Eduard Hermann to Erich Wende in the Ministry of Culture, 22 March 1920, on the qualification of Max Born (GStA PK. Rep. 76 V a, Sekt. 6, Tit. IV, Nr. 1, Bd. XXVII, 73–75, 74–74 back).

Seine ersten Arbeiten behandeln solche Gebiete der theoretischen Physik, zu deren Beherrschung eine sehr weitgehende Kenntnis der mathematischen Hilfsmittel erforderlich ist, darunter insbesondere die Relativitätstheorie in ihrer Anwendung auf bewegte Elektronen. Diese Periode seines Schaffens reicht von 1907 bis 1912. Von da an richtet sich sein Interesse mehr und mehr solchen Problemen zu, deren Erfassung den Besitz eines geschärften Blickes für die Bedürfnisse der Physik zur Voraussetzung hat und zu deren Behandlung die zu überwindenden mathematischen Schwierigkeiten erst in zweiter Linie Anlass geben. Solche Arbeiten der neuen Periode, welche Borns Namen in weiten Kreisen bekannt gemacht haben, sind z. B. seine mit v. Karman veröffentlichter, für die Theorie der spezifischen Wärme wichtiger Satz über die Verteilung der elastischen Eigenschwingungen von Raumgittern und seine zusammen mit Courant verfasste Theorie der Temperaturabhängigkeit der Oberflächenspannung. [...] Sein Hauptinteresse aber ist der Molekulartheorie und dabei besonders der Kristallstruktur zugewandt. Er veröffentlichte Arbeiten über die Drehung der Polarisationsebene in Kristallen und amorphen Körpern, über anisotrope Flüssigkeiten, über die Zerstreuung des Lichtes in Gasen. 1915 fasst er seine Ansichten über die Dynamik der Kristallgitter zusammen und lässt sie in Buchform erscheinen. In diesem Arbeitsgebiet stetig weiter fortschreitend versucht er, über die Natur von elastischen Kräften in Kristallen Klarheit zu gewinnen und es gelingt ihm, sehr wahrscheinlich zu machen, dass jene Kräfte elektrischer Natur sind. Mit Hilfe dieser Hypothese gelangt er zu einer gegenseitigen Verknüpfung vieler Kristalleigenschaften und zu manchen Fragestellungen, welche heute noch einer experimentellen Bearbeitung harren. [...]

Document 19

From an autobiographical sketch by David Hilbert "Über meine Tätigkeit in Göttingen," typescript written after January 1932 (UA-Gö Hilbert Papers, folder 741, Nr. 7, Bl. 24–30, printed in (Reidemeister 1971, 78–82)).

Meine Heimat ist Königsberg, wo ich als Sohn einer alten ostpreußischen Juristenfamilie im Jahre 1862 geboren wurde. (...) Im Jahre 1895, 33 Jahre alt, wurde ich von Felix Klein nach Göttingen berufen, wohl wesentlich auf Grund meiner Arbeiten zur Invariantentheorie, deren zentrale Probleme ich von neuartigen Gesichtspunkten aus in Angriff genommen und gelöst hatte. Die Annahme des Rufes an die Stätte der alten mathematischen Tradition war selbstverständlich. Aber es fiel meiner rau[h] und mir keineswegs leicht, uns sofort in der damals etwas kühlen Göttinger Atmosphäre heimisch zu fühlen. Nicht selten wurden wir mit Kopfschütteln betrachtet, wenn wir uns voller Verständnislosigkeit über die strengen Rangunterschiede hinwegsetzten und zwanglos mit Privatdozenten und gar Studenten verkehrten.

Aber bald hatten wir festen Fuß gefasst. Wir fanden uns mit einer Reihe jüngerer Freunde – unter denen auch Walter Nernst war – zu einer Gruppe Gleichgesinnter zusammen. Mit Felix Klein und seiner Frau verbanden uns trotz der Verschiedenheit der Temperamente bald ein völliges Vertrauen und die gemeinsamen Interessen. Von Anfang an habe ich mich freudig in den Dienst des Kleinschen Ziels gestellt, Göttingen zu einem Zentrum der mathematischen und physikalischen Wissenschaften zu machen.

In allen Fragen der Organisation hatte Klein die unbestrittene und unbedingte Führung; um Dinge der Verwaltung habe ich mich nie gekümmert. Aber wenn es sich um wesentliche Entscheidungen handelte, insbesondere bei Berufungen, bei Schaffung neuer Stellen und dergl., habe ich stets aktiven Anteil genommen. Ich glaube, daß die absolute Einigkeit zwischen Klein und mir sowie den anderen Kollegen, die in allen wichtigeren Fragen bestand, für den Erfolg der Kleinschen Bestrebungen vielfach von großer Bedeutung gewesen ist. Einer dieser Erfolge war die Errichtung einer weiteren mathematischen Professur, in welche mein nächster Freund, Hermann Minkowski, berufen wurde. Es war mir so trotz manchen Verlockungen (mehrfache Rufe nach Berlin, Leipzig, Heidelberg, Bern) nicht schwer, dem Göttinger Wirkungskreise treu zu bleiben und mein wissenschaftliches Lebenswerk hier stetig auszubauen.

Entscheidend für meine Tätigkeit ist die denkbar engste Verbindung zwischen Forschung und Lehre gewesen. (...) Es war mein Grundsatz, in den Vorlesungen und erst recht in den

Seminaren nicht einen eingefahrenen und so glatt wie möglich polierten Wissensstoff, der den Studenten das Führen sauberer Kolleghefte erleichtert, vorzutragen. Ich habe vielmehr immer versucht, die Probleme und Schwierigkeiten zu beleuchten und die Brücke zu den aktuellen Fragen zu schlagen. Nicht selten kam es vor, daß im Verlauf eines Semesters das stoffliche Programm einer höheren Vorlesung wesentlich abgeändert wurde, weil ich Dinge behandeln wollte, die mich gerade als Forscher beschäftigten und die noch keineswegs eine endgültige Gestalt gewonnen hatten. Höhere Vorlesungen dieser Art führten oft zu einer engen Wechselwirkung mit den Zuhörern, welche ihrerseits mit Kritik oder eigenen Gedanken hervortraten. In den Unterhaltungen nach der Vorlesung, bei Spaziergängen und Radfahrten, im Garten, bei Gesellschaften und überhaupt bei jeder sich bietenden Gelegenheit wurden oft solche Diskussionen mit Schülern oder Kollegen fortgesetzt. Das gemeinsame Seminar mit Minkowski und die von Klein gegründete mathematische Gesellschaft mit ihren Nachsitzungen gaben jede Woche neue Anregungen; dieser wissenschaftliche Betrieb wurde in seiner Intensität gesteigert durch die Atmosphäre von Kameradschaft und menschlicher Zusammengehörigkeit, durch die Verbundenheit mit der schönen Natur, durch die Geselligkeit, die den Göttinger mathematischen Kreis unter Führung meiner Frau zu einer großen Familie vereinigte. (...)

Im Jahre 1909 unterbrach ein jäher Schlag den gleichmäßigen Fluß der Dinge. Mein Freund Hermann Minkowski starb ganz plötzlich, und es blieb menschlich und wissenschaftlich für mich eine tiefe Lücke zurück. (...) Kleins umfassende Pläne zur Stabilisierung der Göttinger Einrichtungen schritten fort. ... In dem Moment, wo alles vorbereitet war, das Grundstück und die Mittel bereit standen, brach der Krieg aus. Unser Kreis zerstob in alle Winde. (...) Ich selbst habe mich damals vorzugsweise mit Physik und Relativitätstheorie beschäftigt und nur wenige, zum Teil ausländische, Schüler gehabt.

Bei Kriegsende schienen mit dem Zusammenbruch alle Hoffnungen auf die Vollendung der Kleinschen Pläne begraben zu sein. Die Göttinger Vereinigung löste sich auf. Die mathematischen Einrichtungen, insbesondere das Lesezimmer, befanden sich einem Zustande des Verfalls. Mittel zum Wiederaufbau schienen nicht vorhanden zu sein. Und diese Lage war nicht auf Göttingen beschränkt. So schien z.B. das Drucken von mathematischen und wissenschaftlichen Büchern unmöglich zu sein. In dieser Situation habe ich dann aktiver eingegriffen als ich das früher getan hatte. (...)

Klein und ich waren uns darüber einig, daß alles versucht werden müßte, um einen Wiederaufbau anzubahnen, und daß hierfür die Frage der Neubesetzungen der freigewordenen Plätze von ausschlaggebender Bedeutung sei. In der Physik fiel die Wahl auf Franck und Born. In der Mathematik haben Klein – welcher sich wegen seiner Krankheit zurückzog – und ich die Berufung von Courant bewirkt, welcher die Verpflichtung übernahm, die Kleinschen Pläne aufrechtzuerhalten und fortzusetzen. Zunächst wurde mit Hilfe von Freunden aus der Industrie und zum Teil auch aus Amerika die Inflationszeit überwunden. Schon hierbei bewährten sich die alten Verbindungen, die durch Kleins und meine frühere Lehrtätigkeit in der ganzen Welt angeknüpft worden waren. Dann im Jahre 1927 schien die Zeit reif zu sein, um die alten Aufbaupläne von Klein zu erneuern. Leider hat Klein das Wiederaufblühen seiner Pläne nicht mehr miterlebt. – Unterstützt von Niels und Harald Bohr in Kopenhagen nahm Courant damals Verhandlungen mit der Rockefeller Foundation auf. Das große Ansehen, das sich die Göttinger Mathematik und Physik nach dem Kriege rasch wieder erobert hatte, bereitete den Weg. Nach langwierigen Unterhandlungen bewilligte die Rockefeller Foundation eine sehr erhebliche Summe für die Errichtung des mathematischen Institutes; das preußische Staatsministerium sicherte die Aufrechterhaltung der Mathematik und Physik auf dem entsprechenden Standard zu. Ende 1929 konnte das Göttinger mathematische Institut eingeweiht werden. Meinen 70. Geburtstag habe ich im Januar 1932 im neuen Institut unter dem Eindruck dieses Erfolges feiern können.

References

Abraham, M. (1904). Zur Theorie der Strahlung und des Strahlungsdruckes. *Annalen der Physik, 14*, 236–287.

Barkan, D. K. (1993). The witches' sabbath. The first international Solvay congress in physics. *Science in Context, 6*, 59–82.

Baule, B. (1914). Theoretische Behandlung der Erscheinungen in verdünnten Gasen. *Annalen der Physik.*

Behmann, H. (1918). *Die Antinomie der transfiniten Zahl und ihre Auflsung durch die Theorie von Russell und Whitehead.* Ph.D. thesis, University of Göttingen.

Beller, M. (1983). Matrix theory before Schrödinger: Philosophy, problems, consequences. *Isis, 74*, 469–491.

Beller, M. (1999). *Quantum dialogue.* Chicago: The making of a Revolution.

Bielz, F. (1925). Versuche zur direkten Messung der "mittleren freien Weglänge" von ungeladenen Silberatomen in Stickstoff. *Physikalische Zeitschrift, 32*, 81–102.

Biermann, K.-R. (1988). *Die Mathematik und ihre Dozenten an der Berliner Universität 1810–1933.* Berlin: Akademie.

Blum, A., Jähnert, M., Lehner, C., & Renn, J. (2017). Translation as heuristics: Heisenberg's turn to matrix mechanics. *Studies in History and Philosophy of Modern Physics, 60*, 3–22.

Bohr, N. (1922). Der Bau der Atome und die physikalischen und chemischen Eigenschaften der Elemente. *Zeitschrift für Physik, 9*, 1–67.

Bohr, C. W. *Niels Bohr: Collected Works, 1973–2007.*

Bolza, H. (1913). *Anwendung der Theorie der Integralgleichungen auf die Elektronentheorie und die Theorie der verdünnten Gase.* Ph.D. thesis, Göttingen University.

Born, M. (1913a). Zur kinetischen Theorie der Materie. Einführung zum Kongre in Göttingen (21. bis 26. April). *Naturwissenschaften, 1*, 297–299.

Born, M. (1913b). Die Theorie der Wärmestrahlung und die Quantenhypothese. *Die Naturwissenschaften, 1*, 499–504.

Born, M. (1914a). Elektronentheorie und Relativitätsprinzip. (summer term 1914), lecture notes by Mario Jona, SBB Born Papers II.3, folder 1304.

Born, M., & Langevin u. P., & de Broglie, M., (1914b). La théorie du rayonnement et les quanta. *Physikalische Zeitschrift, 15*, 166–167.

Born, M. (1915). *Dynamik der Kristallgitter.* Teubner.

Born, M. (1918a). Über die Maxwellsche Beziehung zwischen Brechungsindex und Dielektrizitätskonstante und über eine Methode zur Bestimmung der Ionenladung in Kristallen. *Sitzungsberichte der Preussischen Akadademie der Wissenschaften*, 604–613.

Born, M., & Herkner, H. (1918b). *Naturwissenschaften, 6*, 178.

© The Author(s), under exclusive license to Springer Nature Switzerland AG 2019
A. Schirrmacher, *Establishing Quantum Physics in Göttingen*,
SpringerBriefs in History of Science and Technology,
https://doi.org/10.1007/978-3-030-22727-2

Born, M. (1918c). Über die Zerstreuung des Lichtes in H_2O_2 und N_2. *Verhandlungen der Deutschen Physikalischen Gesellschaft, 20,* 16–32.

Born, M. (1919a). Eine thermochemische Anwendung der Gittertheorie. *Verhandlungen der Deutschen Physikalischen Gesellschaft, 21,* 13–24.

Born, M. (1919b). Die Elektronenaffinität der Halogenatome. *Verhandlungen der Deutschen Physikalischen Gesellschaft, 21,* 679–685.

Born, M. (1920a). *Der Aufbau der Materie: Drei Aufsätze über moderne Atomistik und Elektronentheorie.* Springer.

Born, M. (1920b). Eine direkte Messung der freien Weglänge neutraler Atome. *Physikalische Zeitschrift, 21,* 578–581.

Born, M. (1921a). *Über elektrostatische Gitterpotentiale. Zeitschrift für Physik, 7,* 124–140.

Born, M. (1921b). *Die Relativitätstheorie Einsteins und ihre physikalischen Grundlagen.* Springer.

Born, M. (1922a). Hilbert und die Physik. *Die Naturwissenschaften, 10,* 88–93.

Born, M. (1922b). Zur Thermodynamik der Kristallgitter II. *Zeitschrift für Physik, 11,* 327–352.

Born, M. (1923). *Atomtheorie des festen Zustandes.* Teubner.

Born, M. (1924). *Über Quantenmechanik. Zeitschrift für Physik, 26,* 379–395.

Born, M. (1925). *Vorlesungen über Atommechanik. Band I.* Springer.

Born, M. (1926a). Zur Quantenmechanik der Stoßvorgänge. *Zeitschrift für Physik, 37,* 863–867.

Born, M. (1926b). Zur Quantenmechanik der Stoßvorgänge (Vorläufige Mitteilung). *Zeitschrift für Physik, 38,* 803–827.

Born, M. (1975). *Mein Leben.* Die Erinnerungen des Nobelpreisträgers: Nymphenburger Verlag.

Born, M. (1978). *My life.* Recollections of a Nobel laureate: Scribner's.

Born, M., & Bormann, E. (1919a). Zur Gittertheorie der Zinkblende. *Verhandlungen der Deutschen Physikalischen Gesellschaft, 21,* 733–741.

Born, M., & Bormann, E. (1919b). Zur Gittertheorie der Zinkblende II. *Annalen der Physik, 62,* 218–246.

Born, M., & Bormann, E. (1921). Über die interferometrische Methode zur Bestimmung der Dicke dünner Schichten. *Verhandlungen der Deutschen Physikalischen Gesellschaft, 2,* 54.

Born, M., & Brody, E. (1921). Über die Schwingungen eines mechanischen Systems mit endlicher Amplitude und ihre Quantelung. *Physikalische Zeitschrift, 6,* 140–152.

Born, M., & Courant, R. (1913). Zur Theorie des Eötvösschen Gesetzes. *Physikalische Zeitschrift, 14,* 731–740.

Born, M., & Einstein, A. (1969). *Albert Einstein-Hedwig und Max Born: Briefwechsel 1916-1955.* Nymphenburger Verlag.

Born, M., & Franck, J. (1925). *Quantentheorie der Molekelbildung. Zeitschrift für Physik, 31,* 411–429.

Born, M., & Gerlach, W. (1921a). *Elektronenaffinität und Gittertheorie. Zeitschrift für Physik, 5,* 433–441.

Born, M., & Gerlach, W. (1921b). Über die Zerstreuung des Lichtes in Gasen. *Zeitschrift für Physik, 5,* 374–375.

Born, M., & Heisenberg, W. (1923). Über Phasenbeziehungen bei den Bohrschen Modellen von Atomen und Molekülen. *Zeitschrift für Physik, 14,* 44–55.

Born, M., & Jordan, P. (1925a). *Zur quantenmechanik. Zeitschrift für Physik, 34,* 858–888.

Born, M., & Jordan, P. (1925b). Zur Quantentheorie aperiodischer Vorgänge. *Zeitschrift für Physik, 33,* 479–505.

Born, M., & Ladenburg, R. (1911). Über das Verhältnis von Emissions- und Absorptionsvermögen bei stark absorbierenden Körpern. *Physikalische Zeitschrift, 12,* 198–202.

Born, M., & Landé, A. (1918a). Über die absolute Berechnung der Kristalleigenschaften mit Hilfe Bohrscher Atommodelle. *Sitzungsberichte der Preussischen Akadademie der Wissenschaften,* 1048–68.

Born, M., & Landé, A. (1918b). Über die Berechnung der Kompressibilität regulärer Kristalle aus der Gittertheorie. *Verhandlungen der Deutschen Physikalischen Gesellschaft, 20,* 210–216.

Born, M., & Landé, A. (1918c). Kristallgitter und Bohrsche Atommodelle. *Verhandlungen der Deutschen physikalischen Gesellschaft, 20*, 202–209.

Born, M., & Pauli, W. (1922). Über die Quantelung gestörter mechanischer Systeme. *Zeitschrift für Physik, 10*, 137–158.

Born, M., & Stern, O. (1919). Über die Oberflächenenergie der Kristalle und ihren Einfluß auf die Kristallgestalt. *Sitzungsberichte der Preussischen Akadademie der Wissenschaften, 901–913*, 1919.

Born, M., & Wiener, N. (1926). Eine neue Formulierung der Quantengesetze für periodische und nichtperiodische Vorgänge. *Zeitschrift für Physik, 36*, 174–187.

Born, M., Jordan, P., & Nordheim, L. (1925). Zur Theorie der Stoanregung von Atomen und Molekülen. *Naturwissenschaften, 13*, 969–970.

Born, M., Heisenberg, W., & Jordan, P. (1926). Zur Quantenmechanik II. *Zeitschrift für Physik, 35*, 557–615.

Brody, E. (1921). Integralinvarianten und Quantenhypothese. *Zeitschrift für Physik, 6*, 224–228.

Cassidy, D. C. (1992). Uncertainty: The life and science of Werner Heisenberg. *Freeman*.

Castagnetti, G., & Goenner, H. (2004). *Einstein and the Kaiser Wilhelm Institute for Physics (1917-1922). Institutional aims and scientific results (Preprint 261)*. Max Planck Institute for the History of Science.

Condon, E. U. (1926). A theory of intensity distribution in band systems. *Physics Reviews, 28*, 1182–1201.

Corry, L. (1997). David Hilbert and the axiomatization of physics (1894–1905). *Archive for the History of Exact Sciences, 51*, 83–198.

Corry, L. (2004). *David Hilbert and the Axiomatization of Physics (1898–1918)*. From Grundlagen der Geometrie to Grundlagen der Physik: Springer.

Dahms, H.-J. (2002). *Appointment Politics and the Rise of Modern Theoretical Physics at Göttingen*. In Göttingen and the development of the natural sciences: Wallstein.

Dresden, M. (1987). *H. A. Kramers. Between tradition and revolution*. Springer.

Duncan, A., & Janssen, M. (2007). On the verge of Umdeutung in Minnesota: Van Vleck and the correspondence principle. *Archive for the History of Exact Sciences, 61*(553–624), 625–671.

Ebel, W. (Ed.). (1962). *Catalogus Proffessorum Gottingensium*. Vandenhoeck & Ruprecht.

Ebner, F. (2013). *James Frank–Robert Pohl: Briefwechsel 1906–1964*. Deutsches Museum.

Eckert, M. (1993). *Die Atomphysiker*. Eine Geschichte der theoretischen Physik am Beispiel der Sommerfeldschule: Vieweg.

Eckert, M. (2013). *Die Bohr-Sommerfeldsche Atomtheorie*. Springer.

Ehrenfest, P. & Ehrenfest, T. (1911). Begriffliche Grundlagen der statistischen Auffassung in der Mechanik. In *Encyklopädie der mathematischen Wissenschaften Vol. IV/2.II*. Teubner.

Einstein CP. *Altbert Einstein: Collected Papers, 1987–*. Princeton University Press.

Elsasser, W. (1925). Bemerkungen zur Quantenmechanik freier Elektronen. *Naturwissenschaften, 13*, 711.

Eucken, A. (1914). Die Theorie der Strahlung und Quanten. Verhandlungen auf einer von E. Solvay einberufenen Zusammenkunft (30. Oktober bis 3. November 1911); mit einem Anhang über die Entwicklung der Quantentheorie von Herbst 1911 bis zum Sommer, (1913). Abhandlungen der Deutschen Bunsen-Gesellschaft für angewandte physikalische Chemie, *3. Heft, 7*, 1–405.

Ewald, P. P. (1912). *Dispersion und Doppelbrechung von Elektronengittern (Kristallen)*. Ph.D. thesis, Munich University.

Föppl, L. (1912). *Stabile Anordnungen von Elektronen im Atom*. Ph.D. thesis, Göttingen University.

Forman, P. (1970a). Alfred Landé and the anomalous Zeeman effect, 1919–1921. *Historical Studies in the Physical Sciences, 2*, 153–261.

Forman, P. (1970b). Friedrich Paschen. In *Dictionary of scientific biography* (Vol. XII, pp. 613–616). Scribner.

Försterling, K. (1951). Woldemar Voigt zum hundertsten Geburtstage. *Naturwissenschaften, 38*, 217–221.

Franck, J. (1922). Einige aus der Theorie von Klein und Rosseland zu ziehenden Folgerungen über Fluoreszenz, photochemische Prozesse und die Elektronenemission glühender Körper. *Zeitschrift für Physik, 9,* 259–266.

Franck, J. (1924). Zur Frage der Ionisierungsspannung positiver Ionen. *Zeitschrift für Physik, 25,* 312–316.

Frei, G. (Ed.). (1985). *Der Briefwechsel David Hilbert - Felix Klein (1886–1918).* Vandenhoeck & Ruprecht.

Fricke, H. (1974). *150 Jahre Physikalischer Verein Frankfurt a. M.* Physikalischer Verein.

Froben, H. J. (1972). *Aufklärende Artillerie. Geschichte der Beobachtungsabteilungen und selbständigen Beobachtungsbatterien bis 1945.* Schild.

Gerlach, W., & Pauli, O. (1921). Das Gitter des Magnesiumoxids. *Zeitschrift für Physik, 7,* 116–123.

Goldberg, S., & Riecke, E. (1970a). In C. C. Gillispie (Eds.), *Dictionary of scientific biography* (Vol. IV, pp. 445–447). Scribner.

Goldberg, S., & Voigt, W. (1970b). In C. C. Gillispie, (Eds.), *Dictionary of scientific biography,* (Vol. V). Scribner.

Greenspan, N. T. (2005). *The end of the certain world: The life and science of Max Born.* Wiley-VCH.

Heckmann, G. (1924). *Über die Elastizitätskonstanten der Kristalle.* Universität Göttingen.

Heilbron, J. L. (1994). The virtual oscillator as a guide to physics students lost in plato's cave. *Science and Education, 3*(2), 177–188.

Heisenberg, W. (1925). Über quantentheoretische Umdeutung kinematischer und mechanischer Beziehungen. *Zeitschrift für Physik, 33,* 879–893.

Hermann, A., & Stark, J. (1970). *In Dictionary of scientific biography (Vol X).* Scribner.

Hermann, A. (1971). *The genesis of quantum theory.* MIT Press.

Hilbert, D. (1907). Differential- und integralrechunung, 2. teil. (winter term 1906/07), lecture notes, Göttingen, Mathematisches Institut.

Hilbert, D. (1911). *Mechanik der Kontinua (summer team 1911).,* Lecture notes by Erich Hecke Mathematisches Institut: Göttingen.

Hilbert, D. (1912a). Kinetische Gastheorie. (winter term 1911), lecture notes by Erich Hecke, Göttingen, Mathematisches Institut; a slightly altered version with the title Mechanik der Kontinua aufgrund der Atomtheorie can be found in SBB Born Papers III.2 Mappe 1816.

Hilbert, D. (1912b). *Grundzüge einer allgemeinen Theorie der Integralgleichungen.* Teubner.

Hilbert, D. (1912c). Strahlungstheorie. (summer term 1912, Lecture notes by Erich Hecke). Mathematisches Institut Göttingen, 1912.

Hilbert, D. (1914a). Elektromagnetische Schwingungen. (lecture notes winter term 1913/14) by Erich Hecke, Mathematisches Institut Göttingen 1914.

Hilbert, D. (1914b). Statistische Mechanik. (summer term 1914, Lecture notes by Luise Lange), Mathematisches Institut Göttingen.

Hilbert, D. (1915). Die Grundlagen der Physik (Erste Mitteilung). *Nachrichent der Göttinger Gesellschaft der Wissenschaften, 395–407,* 1915.

Hilbert, D. (1935). *Gesammelte Abhandlungen* (Vol. 3). Springer.

Hilbert, D. (1992). *David Hilbert: Natur und mathematisches Erkennen. Vorlesungen gehalten 1919-1920 in Göttingen.* Springer.

Hilbert, D. (2004). *David Hilbert's Lectures on the foundations of geometry, 1891–1902.* Springer.

Hilbert, D. (2009a). *Lectures on the foundations of physics 1915-1927. Relativity, quantum Theory and epistemology.* Springer.

Hilbert, D. (2009b). *David Hilbert's Lectures on the Foundations of Physics 1915-1927. Relativity, Quantum Theory and Epistemology,* chapter Mathematische Grundlagen der Quantentheorie, (pp. 605–707). Springer.

Hilbert, D., von Neumann, J., & Nordheim, L. (1927). Über die Grundlagen der Quantenmechanik. *Mathematische Annalen, 98,* 1–30.

Hoffmann, C. (1994). Wissenschaft und Militär. Das Berliner Psychologische Institut und der Erste Weltkrieg. *Psychologie und Geschichte, 5,* 261–285.

Holl, F. (1996). *Produktion und Distribution wissenschaflicher Literatur. Der Physiker Max Born und sein Verleger Ferdinand Springer 1913-1970*. Buchhändler-Vereinigung.

von Hornbostel, E. M., & Wertheimer, M. (1920). Über die Wahrnehmung der Schallrichtung. *Sitzungsberichte der Preussischen Akadademie der Wissenschaften, 20*, 388–396.

Hund, F. (1923). Theoretische betrachtungen über die ablenkung von freien langsamen elektronen in atomen. *Zeitschrift für Physik, 13*, 241–263.

Hund, F. (1967). *Geschichte der Quantentheorie*. Bibliographisches Institut.

Hund, F. (1987). *Die Geschichte der Göttinger Physik*. Vandenhoeck & Ruprecht.

Gyeong Soon Im. (1995). Formation and development of the Ramsauer effect. *Historical Studies in the Physical and Biological Sciences, 25*(2), 269–300.

Gyeong Soon Im. (1996). Experimental constraints on formal quantum mechanics: The emergence of Born's quantum theory of collision processes in Göttingen 1924–1927. *Achive for the History of Exact Sciences, 50*, 73–101.

Instituts Solvay. (1921). Conseil de physique. *La structure de la matiere. Rapports et discussion du Conseil de physique tenu a Bruxelles du 27 au 31 Octobre 1913*. Gauthier-Villars et cie.

Ioffe, A. (1967). *Begegnungen mit Physikern*. Teubner.

Jähnert, M. (2015). Practising the correspondence principle in the old quantum theory: Frank, Hund and the Ramsauer effect. In F. Aaserud, & H. Kragh, (Eds.), *One Hundred Years of the Bohr Atom* (pp. 200–216). Royal Danish Academy of Sciences.

Jähnert, M. (2019). *Practicing the correspondence principle in the old quantum theory-A transformation through implementation*. Springer.

Jammer, M. (1966). *The conceptual development of quantum mechanics*. McGraw-Hill.

Jb. DMV. *Jahresberichte der Deutschen Mathemativer-Vereinigung*.

Joas, C., & Lehner, C. (2009). The classical roots of wave mechanics: Schrödinger's transformations of the optical-mechanical analogy. *Studies in History and Philosophy of Modern Physics, 40*, 338–351.

Jungnickel, C., & McCormmach, R. (1986). *The Intellectual mastery of nature: Theoretical Physics from Ohm to Einstein. Vol. 2, The now mighty theoretical physics 1870 to 1925*. University of Chicago Press.

Kaiser, W., & Ewald, P. P. (1970) In C. C. Gillispie (Eds.), *Dictionary of scientific biography*, (Vol. XI, pp. 445–447). Scribner.

Kallmann, H., & Reiche, F. (1921). Über den Durchgang bewegter Moleküle durch inhomogene Kraftfelder. *Zeitschrift fr Physik, 6*, 352–375.

Kangro, H. (1969). Das Paschen-Wiensche Strahlungsgesetz und seine Abänderung duch Max Planck. *Physikalische Blätter, 25*, 216–220.

Kayser, H. (1996). Erinnerungen aus meinem Leben [1936]. *Institut für Geschichte der*,. Naturwissenschaften.

Klein, F. (1908). Die Göttinger Vereinigung zur Förderung der angewandten Physik und Mathematik. *Internationale Wochenschrift für Wissenschaft, Kunst und Technik, 2*, 519–534.

Klein, M. J. (1970). *Paul Ehrenfest: Vol. 1, The making of a theoretical physicist*. North-Holland.

Klein, O., & Rosseland, S. (1921). Über Zusammenstöße zwischen Atomen und freien Elektronen. *Zeitschrift für Physik, 4*, 46–51.

Kneser, H. (1921). Untersuchungen zur Quantenlehre. *Mathematische Annalen, 76*, 228–232.

Kojevnikov, A. (forthcoming). *"Knabenphysik": Itinerant Postdocs and the Invention of Quantum Mechanics*.

Kratzer, A. (1920). Die ultraroten Rotationsspektren der Halogenwasserstoffe. *Zeitschrift für Physik, 31*, 681–709.

Kuhn, T. S. (1978). *Black-body theory and the Quantum Discontinuity, 1894-1912*. University Of Chicago Press.

Ladenburg, R. (1921). Die quantentheoretische Deutung der Zahl der Dispersionselektronen. *Zeitschrift für Physik, 4*, 451–468.

Landé, A. (1914). *Zur Methode der Eigenschwinungen in der Quantentheorie*. Ph.D. thesis, Munich University.

Landé, A. (1920). *Störungstheorie des Heliumatoms. Physikalische Zeitschrift, 21*, 114–122.

Landé, A. (1920). Würfelatome, peridisches System und Molekülbindung. *Zeitschrift für Physik, 2*, 380–404.

Lemmerich, J. (2007). *Aufrecht im Sturm der Zeit. Der Physiker James Franck, 1882–1964.* GNT-Verlag.

Lertes, P. (1920). *Untersuchungen über Rotationen von dielektrischen Flüssigkeiten im elektrostatischen Drehfeld.* Ph.D. thesis, Frankfurt University.

Lertes, P. (1921a). Der Dipolrotationseffekt bei dielektrischen Flüssigkeiten. *Zeitschrift für Physik, 6*, 56–68.

Lertes, P. (1921b). Untersuchungen über Rotationen von dielektrischen Flüssigkeiten im elektrostatischen Drehfeld. *Zeitschrift für Physik, 4*, 315–336.

Lorentz, H. A. (1910). Alte und neue Fragen der Physik (elaboration of talks of H. A. Lorentz by Max Born). *Physikalische Zeitschrift, 11*, 1234–57.

Lorey, W. (1916). *Das Studium der Mathematik an den deutschen Universitäten seit Anfang des 19. Jahrhunderts:* Teubner.

Lustiger, A. (1994). *Jüdische Stiftungen in Frankfurt am Main.* Waldemar Kramer.

Madelung, E. (1918). Das elektrische Feld in Systemen von regelmässig angeordneten Punktladungen. *Zeitschrift für Physik, 19*, 542–553.

McCormmach, R. (1982). *Night thoughts of a classical physicist.* Harvard University Press.

McGuinness, B. (Ed.). (1987). *Unified Science.* The Vienna Circle Monograph Series: Springer.

Mehra, J. (1975). *The Solvay conferences on physics-aspects of the development of physics since 1911.* Dordrecht:

Mehra, J., & Rechenberg, H. (1982). *The historical development of quantum theory, Vol. 3: The formulation of matrix mechanics and its modifications 1925-1926.* Springer, New York.

Mie, G. (1917a). Die Einsteinsche Gravitationstheorie und das Problem der Materie (I). *Physikalische Zeitschrift, 18*, 551–556.

Mie, G. (1917b). Die Einsteinsche Gravitationstheorie und das Problem der Materie (II). *Physikalische Zeitschrift, 18*, 574–580.

Mie, G. (1917c). Die Einsteinsche Gravitationstheorie und das Problem der Materie (III). *Physikalische Zeitschrift, 18*, 596–602.

Minkowski, H. (1973). *Briefe an David Hilbert.* Springer.

Minkowski, R., & Sponer, H. (1923). Über die freie Weglänge langsamer Elektronen in Gasen. *Zeitschrift für Physik, 15*, 399–408.

Minkowski, R., & Sponer, H. (1924). Über den Durchgang von Elektronen durch Atome. *Ergebnisse der exakten Wissenschaften, 3*, 67–85.

Nachr. GWG. *Nachrichenten der Göttinger Gesellschaft der Wissenschaften.*

Nordheim, L. (1923a). Zur Behandlung entarteter Systeme in der Störungsrechnung. *Zeitschrift für Physik, 17*, 316–330.

Nordheim, L. (1923b). Zur Quantentheorie des Wasserstoffmolekülions. *Zeitschrift für Physik, 19*, 69–93.

Nordheim, L. (1926). Zur Theorie der Anregung von Atomen durch Stöße. *Zeitschrift für Physik, 36*, 496–539.

Paschen, F. (1916). Bohrs Heliumlinien. *Annalen der Physik, 50*, 901–940.

Pauli, W. (1922). Über das Modell des Wasserstoffmolekülions. *Annalen der Physik, 68*, 177–240.

Peckhaus, V. (1990). *Hilbertprogramm und kritische Philosophie.* Vandenhoeck und Ruprecht: Das Göttingen Modell interdisziplinärer Zusammenarbeit zwischen Mathematik und Philosophie.

Planck, M. (1906). *Vorlesungen über die Theorie der Wärmestrahlung.* Johann Ambrosiums Barth Leipzig.

Planck, M., Debye, P., Nernst, W., Smoluchowski, M. V., Sommerfeld, A., & Lorentz, H. A. (1914). *Vorträge über die kinetische Theorie der Materie und der Elektrizität (Gehalten in Göttingen auf Einladung der Wolfskehlstiftung).* Teubner.

Planck, M., et al. (1913). Vorbericht für den von der Kommission der Wolfskehlstiftung veranstalteten Zyklus von Vorträgen über die kinetische Theorie der Materie. *Nachr. Ges. Wiss. Göttingen, 1913*, 137–156.

Poincare, H. (1912). Sur la thorie des quanta. *Journal de physique, 2*, 5–34.

Ramsauer, C. (1921). Über den Wirkungsquerschnitt der Gasmoleküle gegenüber langsamen Elektronen. *Annalen der Physik, 369*, 513–540.

Reiche, F. (1913a). Ueber die Emission. *Absorption und Intensitätsverteilung von,*. Spektrallinien.

Reiche, F. (1913b). *Die Quantentheorie. Naturwissenschaften, 1*, 568–572.

Reiche, F. (1921). *Die Quantentheorie. Ihr Ursprung und ihre Entwicklung.*

Reiche, F., & Ladenburg, R. (1912). Über selktive Absorption. Jahresbericht der Schlesischen Gesellschaft für vaterländische Cultur. *Naturwissenschaftliche Sektion, 1912*, 1–20.

Reid, C. (1970). *Hilbert.* Springer.

Reidemeister, K. (1971). *Hilbert.* Gedenkband: Springer.

Renn, J. (2000). Challenges from the Past. Innovative Structures for Science and the Contribution of the History of Science. In Max-Planck-Gesellschaft zur Förderung der Wissenschaften, editor, *Innovative Structures in Basic Research, Ringberg-Symposium, 4-7 October 2000.*

Riecke, E. (1915). Bohrs Theorie der Serienspektren von Wasserstoff und Helium. *Physikalischer Zeitschrift, 16*, 222–227.

Ritz, W. (1903). Zur Theorie der Serienspektren. *Annalen der Physik, 12*, 264–310.

Ritz, W. (1908). Über ein neues Gesetz der Serienspekten. *Physikalische Zeitschrift, 9*, 521–529.

Rowe, D. (1985). Jewish mathematicians in göttingen in the era of felix klein. *ISIS, 77*, 442–449.

Rowe, D. (1989). Klein, Hilbert and the Göttingen mathematical tradition. *Osiris, 5*, 186–213.

Rowe, D. (1992). *Felix Klein, David Hilbert, and the Gottingen mathematical tradition.* City University of New York.

Rowe, D. (2001). Einstein meets Hilbert: At the crossroads of physics and mathematics. *Physics in Perspective, 3*, 379–424.

Rowe, D. (2009). Mathematik an der Göttinger Universität. In *Jüdische Mathematiker in der deutschsprachigen akademischen Kultur.* Springer.

Runge, C. (1907). Über die Zerlegung von Spektrallinien im magnetischen Felde. *Physikalische Zeitschrift, 8*, 232–237.

Runge, C. (1920). Woldemar Voigt. *Nachrichent der Göttinger Gesellschaft der Wissenschaften, 46–52*, 1920.

Runge, I. (1949). *Carl Runge und sein wissenschaftliches Werk.* Vandenhoeck & Ruprecht.

Sauer, T. (1999). The relativity of discovery: Hilbert's first note on the foundations of physics. *Archive for the History of Exact Sciences, 53*, 529–575.

Sauer, T., & Majer, U. (2005). *Hilbert's "world equations" and his vision of a unified science.* In The Universe of General Relativity: Springer.

Sauer, T., & Majer, U. (2009). *David Hilbert's Lectures on the Foundations of Physics 1915–1927.* Relativity: Quantum Theory and Epistemology. Springer.

Schellenberg, K. (1915). Anwendungen der Integralgleichungen auf die Theorie der Elektrolyse. *Annalen der Physik, 47*, 81–127.

Schirrmacher, A. (2000). Die Rolle materieller Ressourcen in der Wissenschaftsgeschichte. Philipp Lenard und die Apparate. In C. Meinel, (Ed.), *Instrument-Experiment: Historische Studien* (pp. 386–395). GNT.

Schirrmacher, A. (2002). The Establishment of Quantum Physics in Göttingen 1900-24. Conceptional Preconditions–Resources–Research Politics. In H. Kragh, P. Marage, & G. Vanpaemel, (Eds.), *History of Modern Physics. Proceedings of the XXth International Congress of History of Science (Lige 20–26 July, 1997)* (pp. 295–309). Brepols.

Schirrmacher, A. (2003). Planting in his neighbor's garden: David Hilbert and early Göttingen quantum physics. *Physics in Perspective, 5*, 4–20.

Schirrmacher, A. (2009). Von der Geschossbahn zum Atomorbital? Möglichkeiten der Mobilisierung von Kriegs- und Grundlagenforschung füreinander in Frankreich, Grossbritannien und Deutschland, 1914–1924. In M. Berg, J. Thiel, & P. Walther, (Eds.), *Mit Feder und Schwert. Militär und Wissenschaft–Wissenschaftler und Krieg*, (pp. 155–175). Steiner.

Schirrmacher, A. (2012). Ein physikalisches Konzil. Wie die Solvay-Konferenz und das Solvay-Institut von hundert Jahren nicht nur der Quantenmechanik zum Durchbruch verhalfen. *Physik Journal*, *11*(1), 39–42.

Schirrmacher, A. (2015). Who made quantum theory popular with physicists and beyond? the solvay model, a new center for quantum physics, and science communication. *European Physical Journal Special Topics*, *224*, 2113–2125.

Schirrmacher, A. (2016). Sounds and repercussions of war: Mobilization, invention and conversion of First World War science in Britain. *France and Germany. History and Technology*, *32*, 269–292.

Schmidt-Böcking, H., & Reich, K. (2011). *Otto Stern: Physiker, Querdenker*. Nobelpreisträger: Societäts-Verlag.

Schrödinger, E. (1922). Über die spezifische Wärme fester Körper bei hoher Temperatur und über die Quantelung von Schwingungen endlicher Amplitude. *Zeitschrift für Physik*, *11*, 170–176.

Segre, E. (1973). *Biographical Memoir*, chapter Otto Stern, 1888-1969, (pp. 215–236). National Academy of Sciences.

Sigurdsson, S. (1991). *Hermann Weyl, mathematics and physics, 1900-1927*. Ph.D. thesis, Harvard University.

Solvay, E. (1911). *Sur l'etablissement des principes fondamentaux de la gravito-materialitique*. Bothy.

Sommerfeld, A., & Born, M. (1915). Dynamik der Kristallgitter. *Naturwissenschaften*, *3*, 669–670.

Sommerfeld, A. (1916a). Die Quantentheorie der Spektrallinien und die letzte Arbeit von Karl Schwarzschild. *Die Umschau*, *20*, 941–946.

Sommerfeld, A. (1916b). *Eduard riecke*. In Jahrbuch der Königlich Bayerischen Akademie der Wissenschaften: Beck.

Sommerfeld, A. (1917). Die Drudesche Dispersionstheorie vom Standpunkte des Bohrschen Modells und die Konstitution von H_2, O_2 und N_2. *Annalen der Physik*, *53*, 497–549.

Sommerfeld, A. (1922). *Atombau und Spektrallinien*. Vieweg, 3. rev. edition.

Sommerfeld, A. (2000). *Wissenschaftlicher Briefwechsel* (Vol. 1, pp. 1892–1918). GNT-Verlag.

Sommerfeld, A. (2004). *Wissenschaftlicher Briefwechsel* (Vol. 2, pp. 1919–1951). GNT-Verlag.

Sponer, H. (1923). Über freie Weglängen langsamer Elektronen in Edelgasen. *Zeitschrift für Physik*, *18*, 249–257.

Staley, R. (1992). *Max Born and the German physics community*. Cambridg: The education of a physicist.

Stark, J. (1987). *Erinnerungen eines deutschen Naturforschers*. Bionomica.

Stern, O. (1920). Eine direkte Messung der thermischen Molekulargeschwindigkeit. *Zeitschrift für Physik*, *2*, 49–56.

Stern, O. (1921). Ein Weg zur experimentellen Prüfung der Richtungsquantelung im Magnetfeld. *Zeitschrift für Physik*, *7*, 249–253.

Stern, O., & Gerlach, W. (1922). Der experimentelle Nachweis der Richtungsquantelung. *Zeitschrift für Physik*, *9*, 349–352.

Tobies, R. (1981). *Felix Klein. Unter Mitw. von Fritz König. Biographien hervorragender Naturwissenschaftler, Techniker und Mediziner; 5*. Teubner.

Tobies, R. (1991). Wissenschaftliche Schwerpunktbildung: der Ausbau Göttingens zum Zentrum der Mathematik und Naturwissenschaften. In *Wissenschaftsgeschichte und Wissenschaftspolitik im Industrezeitalter: Das "System Althoff" in historischerr Perspektive*. Edition Bildung und Wissenschaft, Verlag A. Lax.

Tobies, R. (1994). Albert Einstein und Felix Klein. *Naturwissenschaftliche Rundschau*, *47*, 345–352.

Toeplitz, O., & Hellinger, E. (1906). Grundlagen für eine Theorie der unendlichen Matrizen. *Nachrichent der Göttinger Gesellschaft der Wissenschaften*, *351–355*, 1906.

Toeplitz, O., & Hellinger, E. (1910). Grundlagen für eine Theorie der unendlichen Matrizen. *Mathematische Annalen, 69*, 289–330.

Toeplitz, O. & Hellinger, E. (1927). Integralgleichungen und Gleichungen mit unendlich vielen Unbekannten. In *Enzyklopädie der mathematischen Wissenschaften Vol. II/3*. Teubner.

Tollmien, C. (1991). Die Habilitation von Emmy Noether an der Universität Göttingen. NTM: Zeitschrift fur Geschichte der Naturwissenschaft. *Technik und Medizin, 28*, 1–32.

Voigt, W. (1906). Rede bei der Eröffnungsfeier. In *Die Physikalischen Institute der Universität Göttingen, Göttinger Vereinigung zur Förderung der angewandten Physik und Mathematik*. Göttinger Vereinigung.

Voigt, W. (1912). *Physikalische Forschung und Lehre in Deutschland während der letzen hundert Jahre*. Vandenhoeck & Ruprecht.

Voigt, W. (1915a). Stark's "Elektrische Spektralanalyse chemischer Atome" (review). *Göttingscher gelehrter Anzeiger, 7*(8), 500–504.

Voigt, W. (1915b). Gustav Rümelin und Werner Planck. *Physikalische Zeitschrift, 15*, 65–68.

Voigt, W. (1915c). Nachruf auf Eduard Riecke. In *Chronik der Georg-August-Universität zu Göttingen für das Rechnungsjahr 1915*. Göttingen University.

Voigt, W. (1918). Struktur und Elastizitätstheorie regulärer Kristalle. *Nachrichtender Göttinger Gesellschaft der Wissenschaften, 1918*(1–32), 33–50.

Voigt, W., & Hansen, H. M. (1912). Das neue Gitterspektroskop des Göttinger Instituts und seine Verwendung zur Beobachtung magnetischer Doppelbrechung im Gebiete der Absorptionslinien. *Physikalische Zeitschrift, 13*, 217–224.

vom Brocke, B. (1980). Hochschul- und Wissenschaftspolitik in Preußen und im Deutschen Kaiserreich 1882-1907: Das System "Althoff". In P. Baumgart, (Ed.), *Bildungspolitik in Preuen zur Zeit des Kaiserreichs*, (pp. 9–118). Klett-Cotta.

vom Brocke, B. (1981). Der deutsch-amerikanische Professorenaustausch. Preußische Wissenschaftspolitik, internationale Wissenschaftsbeziehungen und die Anfänge der deutschen auswärtigen Kulturpolitik vor dem Ersten Weltkrieg. *Zeitschrift für Kulturaustausch, 31*, 128–182.

Weigt, H. (1921). *Die elektrischen Momente des CO und CO2 Moleküls*. Ph.D. thesis, Göttingen University.

Weyl, H. (1912a). Über das Spektrum der Hohlraumstrahlung. *Journal für die reine und angewandte Mathematik (Crelles Journal), 141*, 163–181.

Weyl, H. (1912b). Das asymptotische Verteilungsgesetz der Eigenwerte linearer partieller Differentialgleichungen (mit einer Anwendung auf die Theorie der Hohlraumstrahlung). *Mathematische Annalen, 71*, 441–479.

Weyl, H. (1913). Über die Randwertaufgaben der Strahlungstheorie und asymptotische Spektralgesetze. *Journal für die reine und angewandte Mathematik, 143*, 26.

Wiener, N. (1926). Operational calculus. *Mathematische Annalen, 85*, 557–584.

Wolff, S. L. (1996). Woldemar Voigt und Pieter Zeeman - eine wissenschaftliche Freundschaft. In D. Hoffmann, F. Bevilacqua, & R. H. Stuewer (Eds.), *The emergence of modern physics*.

Name Index

Printed in the United States
By Bookmasters